UNIFIED FIELD FUSION PHYSICS

NORDHAMMER PUBLISHERS

UNIFIED FIELD FUSION PHYSICS
By Greg Castle
Nordhammer Publishers

ISBN 978-1-329-79062-9

UNIFIED FIELD FUSION PHYSICS

Octalinear

dimension:

project:

Ver: 2

5 4 3 2 1

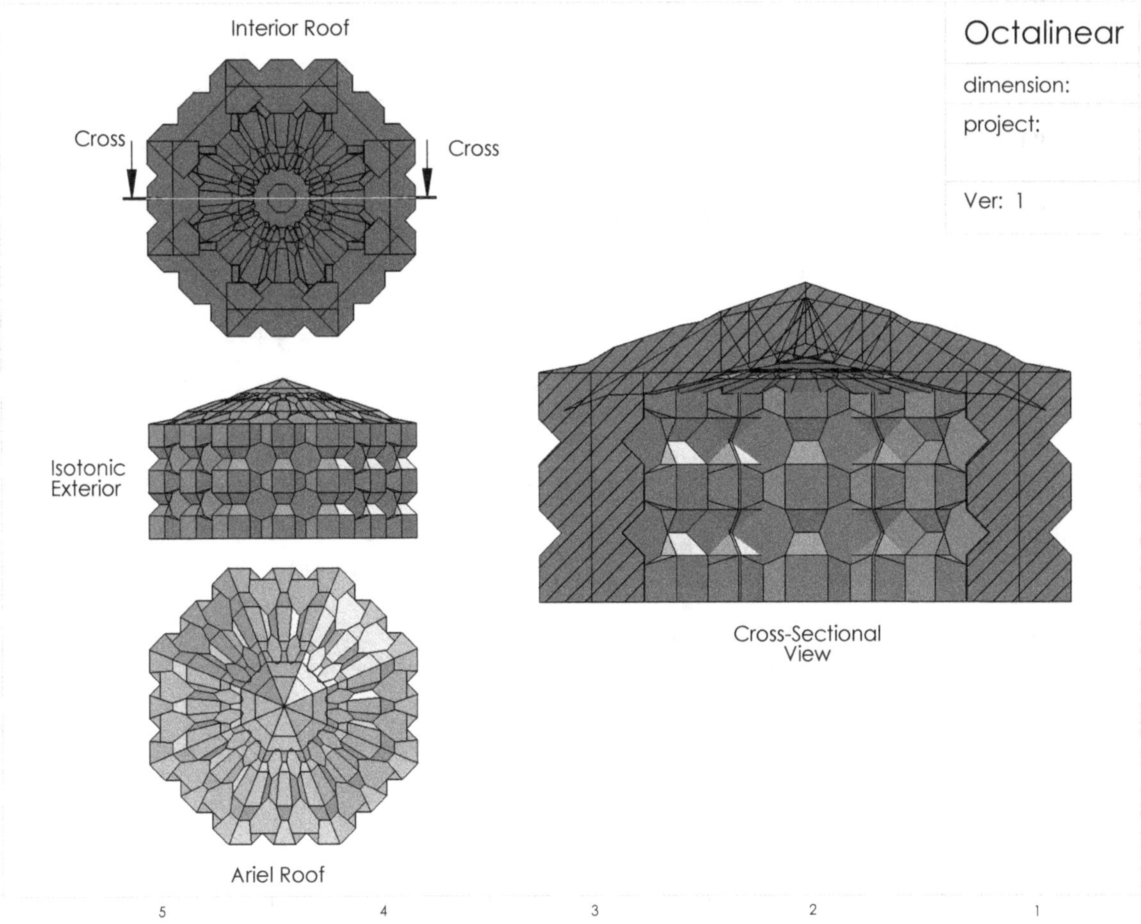
Interior Roof
Cross
Cross
Octalinear
dimension:
project:
Ver: 1
Isotonic Exterior
Cross-Sectional View
Ariel Roof
5
4
3
2
1

1). ~~$E=MC^2$~~ $=A=B \times M \times ACC \times C=$ SIMULTANEITY

2). @AXIS (C) $\equiv \left(S=\frac{\infty}{\infty}=\theta C \sqrt{\frac{\Omega_0}{0^\circ}}\right)=(A^2+B^2)$

3). A/G∞/ACC/@/OR A@ 90°@ OR∞ ACC $=\infty$

4). $136<|\alpha|<137=P=\frac{MV^2}{2}$ @∞ACC $=\alpha$

5). $\alpha<136=\infty$ACC $=\alpha=\infty=186{,}000$

6). MP $<137=134{,}000$ MPS $=P=\frac{MV^2}{2}$

7). $\equiv$(QUANTUM LEAP) $+52{,}000$

8). MPH [QUANTUM LEAP OCCURS

9). IN SIMULTANEITY@AXIS(C)

10). $=136<186{,}000$ MPS$<=$

11). FACTOR) (POSITIVE POWER

12). SIMULTANAEITY@AXIS(C)=

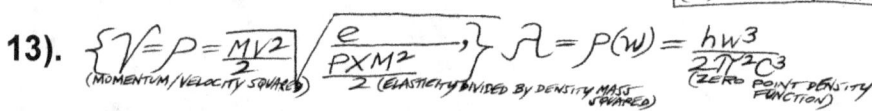

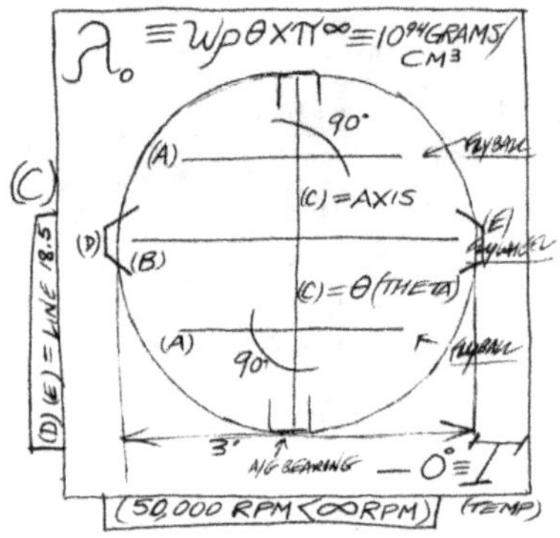

13). $\left\{ \mathcal{V}=P=\frac{MV^2}{2} \sqrt{\frac{e}{\frac{P \times M^2}{2}}}, \right\} \Omega=\rho(w)=\frac{hw^3}{2\pi^2 C^3}$

(MOMENTUM/VELOCITY SQUARED) (ELASTICITY DIVIDED BY DENSITY MASS SQUARED) (ZERO POINT DENSITY FUNCTION)

14). $<\infty$ACC $//C_1=C_0\left(\frac{wp-}{wp+}\right)^2//\Omega_0=C_0=\epsilon_0\frac{S}{D}$

(RESONANT WAVE AMPLIFICATION) (ELECTRO VACUUM CAPACITANCE)

15). $=\infty$ACC $=10^{94}$ GRAMS/CM3 $=136<|\alpha|<137=\infty$

(VALENCE POTENTIAL@ ZERO POINT ENERGY) (INDUCTION AMALGUM FORMULA)

16). $\equiv\left\{\mathcal{V}=f.\Omega<\frac{\Omega}{\infty}=\frac{\mathcal{V}\infty}{f\infty}\right\}\infty=$ {[134,000 MP <186,000 MPS]

(LIGHT WARP ACCELERATION FORMULA) (INF SPACE CREATED BY INF VEL/FREQ) (<=52,000 MPH QUANTUM LEAP) +36% POWER FACTOR

17). $\equiv \rho(w)\frac{hw^3}{2\pi^2 C^3}<\frac{\infty}{ACC}\equiv\sum\overline{136<|\alpha|<137}$

18). (BOYER/LORENTZ INVARIANT) (GENERAL RELATIVITY)

[ZERO POINT DENSITY FUNCTION]

8.5) $\phi \cdot \rho V^2=\int_{10^{12}}^{10^{94}\text{ GRAMS/CM3}}\frac{\mathcal{I}}{(\Omega-\mathcal{I})}\times\frac{\theta}{I^\infty}\cdot\frac{E_0}{(\pi^2.\infty ACC)}<a_0\cdot\oplus$ ←(MAGNETIC VACUUM BUCKING EQUATION) 271.B.3

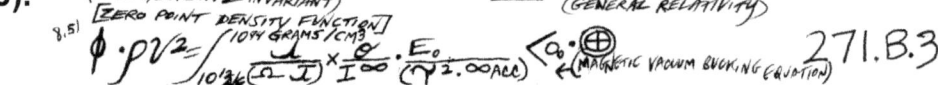

19). $= \{ w_p \theta \sqrt{\frac{NQ^2}{\epsilon_0 m}} \chi \Omega_0 = L = \frac{dm}{SNQ^2} /$ (CARRIER CHARGE/CARRIER MASS) (THETA WAVE PLASMA FREQ) (VACUUM) (EQUIVELANT INDUCTION) (ELECTRO SEPERATION) (ELECTRO AREA)

20). $C_0 = \epsilon_0 \frac{S}{D} = \infty = \theta w = \sqrt{\frac{1}{L(C_0 + C_1)}} =$ (THETA WAVE) (CIRCUIT RESONATION →

21). $\sqrt{\frac{SNQ^2}{dm(C_0+C_1)}} = \sqrt{\frac{NQ^2}{\epsilon_0 m (1 + C_1/C_0)}}$ (CIRCUIT RESONATION) →

22). $= \frac{w_p - (\theta)}{\sqrt{1 + C_1 C_0}} \} = \Omega_0 \; \frac{1}{T} \equiv 0^\circ \sqrt{\frac{P = \frac{MV^2}{2} / 10^{94}\,GRAMS/CM3}{10^{12} = A/G \equiv LEVITATION}}$ (ANTI-GRAVITY FUNCTION)

23). $\equiv w_p \theta \times \pi^\infty \equiv 10^{94}\,GRAMS/CM3 \cong$ (FLYWHEEL ACCELERATION FORMULA)

24). $(\sum \frac{1}{2} h\nu) = \sum_a \partial^2 Q / \partial x a^2 = \mu^2 Q$ (INFINITE ZERO POINT ENERGY) (SCHRÖNDINGER SCALAR WAVE EQUATION)

25). $\sum = evr/c = mv^2/H = r = mcv/eH$ (AVERAGE MAGNETIC MOMENT) (CHARGE/MASS PARTICLE)

26). $\int_\infty^t \frac{\alpha}{\infty} \int_{0^\circ}^\infty = \psi \Omega \circ \psi (X) \equiv \sum_{\alpha 136}^{137} \varphi \xi \cdot \psi$ (CONT) (27)

27). $136 < |\alpha| < 137$ (27) $\sum_{\alpha 136}^{137} \varphi \xi \cdot \psi \Omega^3 \int_{0^\circ}^\infty = \psi \Omega^4 \cdot \pi \infty \frac{MV^2}{2} \Omega_3 \sqrt{10^{12} A/G \equiv LEVITATION / 10^{94} GRAMS/CM3}$ (AMALGUM EQUATION/QUANTUM LEAP) (FREE RADICALIZED VACUUM STATE)

28). $\sum \int_{(\phi \Delta \frac{33^\circ}{180^\circ})} \equiv \Delta^4 \equiv [\partial m_\infty = \frac{2}{4} = \frac{-0^\circ}{T}] \int \Omega_4 \equiv (\Delta\, 33^\circ 3 \cdot 2) \equiv \frac{\infty}{\pi}$ (OPEN DELTA) (4 DIMENSIONAL STATE) SC (E/M MASS) (ANNIHILATION OF THERMODYNAMICS) SCALAR WAVE INTEGRAL (PRIME MOVER/TORROID ARRAYS) (OPEN DELTA INDUCTION ARRAY) (OPEN DELTA FLYWHEEL TORROID IN CADUCEAN COIL)

29). $\int_E^M \} \mp \Omega_0 \sqrt{\frac{\dot{\gamma} \mu \nu}{E = h\nu / P = h/\Omega}} \cong \rho (T, \Omega \infty)$ (ANNIHILATION OF LAMBDA THROUGH VELOCITY) (ENERGY/MASS INTEGRAL FUNCTION) (ZERO POINT ENERGY) (PLANCKS' CONSTANT) / (DE BROGIE'S CONSTANT)

30). $\sum_{N=0}^\infty e \Omega_N Z_N (T) = P = \frac{MV^2}{2} \sqrt{\frac{\int_\infty^t \frac{\alpha}{\infty} \int_{0^\circ}^\infty}{\Omega_0 / T_0 / (\Omega) 136 < |\alpha| < 137 (A)}}$ (ANTI-GRAVITY/ANTI-MATTER UNIFIED FIELD EQUATION)

31). $\cong \sum_{\alpha 136}^{137} \pi^\infty \xi \sqrt{\frac{T(c) \doteq 136 < |\alpha| < 137}{G(10^{94}/CM3) \cdot P = \frac{MV^2}{2}}} = \int_{\Omega_4}^{-0^\circ T} = \Omega_0 = \infty$ (PAGE NUMBER 271.B4)

32). (TIME DISPLACEMENT THEORY)

33). $E^{\infty} \equiv$ (NUCLEAR FUSION) $\circledcirc \phi \frac{\infty}{\pi} \left[\tau^{60°} \times 3 = \Omega^{\infty} \right]$ (NUCLEAR FUSION CHOP WAVE THREE PHASE ROTARY CORE)

34). $= \frac{1}{P} dp = -d\Phi ; \Phi = V - \frac{1}{3} \frac{\Omega^2}{(\tau^3 \infty_{ACC})} W^2 :$

(UNIFORM MICRO SOLAR ROTATING CONFIGURATION)

35). $\triangle^{\cdot 33°} = \Phi = \Phi(A)$, (SURFACE CONSTANT FOR ISOBARIC – ISOPYCNIC SURFACES ISOMETRIC ROTATING SYSTEM W/ CONSTANT CENTER OF MASS)

36). $r(a,\theta) = a\left[1 - \sum_{n=1}^{\infty} \epsilon_{2n}(a) P_{2n}(\cos\theta)\right], \equiv$

(UNIFORMLY ROTATING BODY FOR MICRO-SOLAR FUSION EVENT)

37). SIMULTANAETY @ AXIS C $\left(S = \frac{\infty}{\infty} = \theta^{c} \sqrt{\frac{\Omega \cdot}{0°}}\right)$

$\equiv E^{\infty}$ (PERPETUAL NUCLEAR FUSION)

36). LINE (36) (THE CLAIRAUT-LEGENDRE EXPANSION)

$$r(a,\theta) = a\left[1 - \sum_{n=1}^{\infty} \epsilon_{2n}(a) P_{2n}(\cos\theta)\right], \equiv$$

(UNIFORMLY ROTATING BODY FOR MICRO-SOLAR FUSION EVENT)

SPACIO TEMPORAL WARP EVENT

$$\pi \cdot \Omega^{2} \cdot A/G \left(P = \frac{MV^2}{2}\right) = \pi^{\infty} \left(T \cdot \Omega^{\infty}\right) \equiv 0° \equiv$$

$$C = 60° \cdot 3 \triangle = 1.41 \left(C\sqrt{2}\right)$$

QUANTUM LEAP

30).

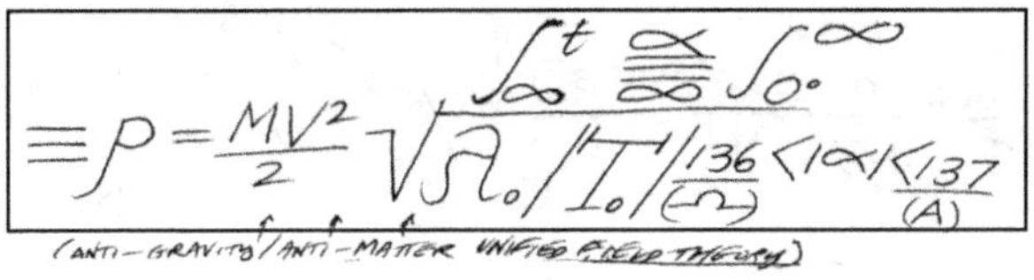

22).

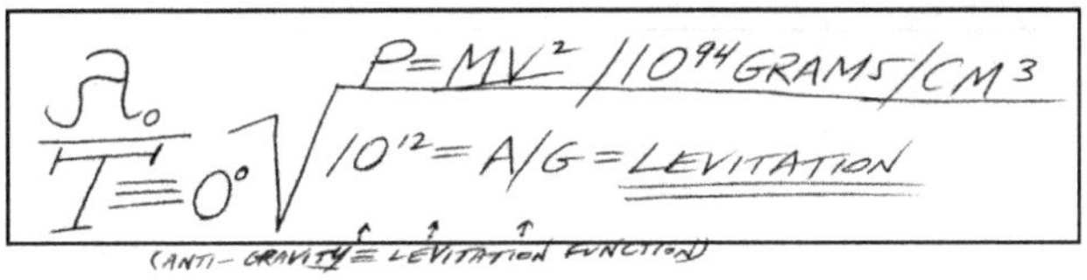

2).

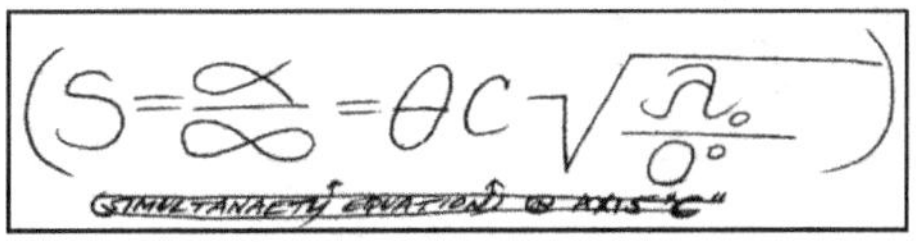

28).

23).

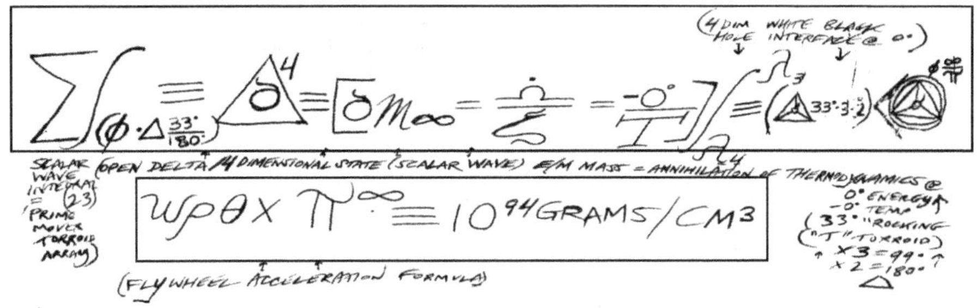

17).

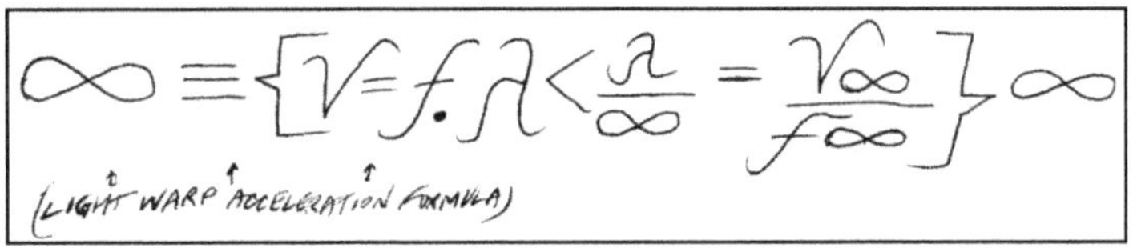

27.5).

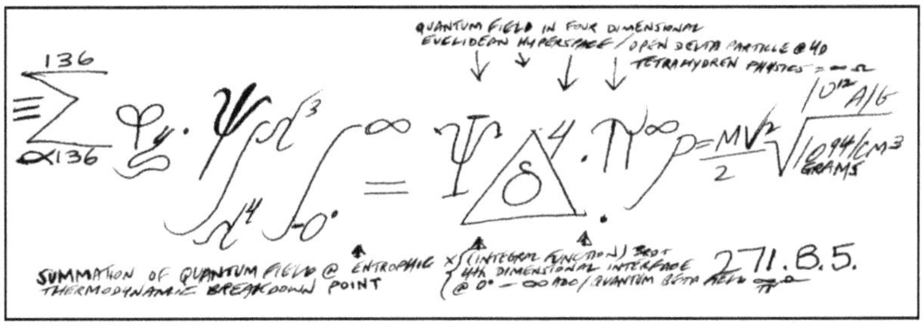

31).

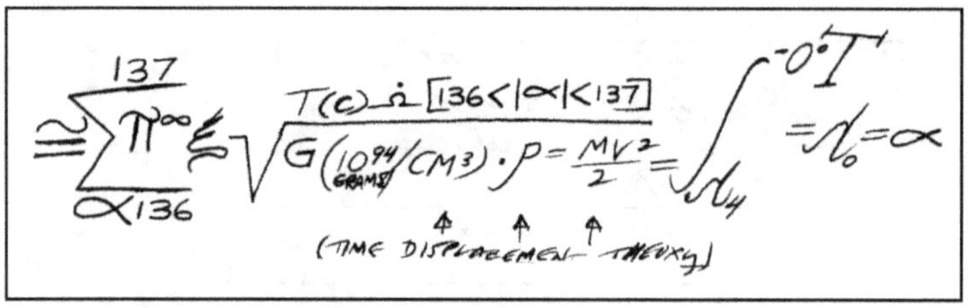
(TIME DISPLACEMENT THEORY)

32).

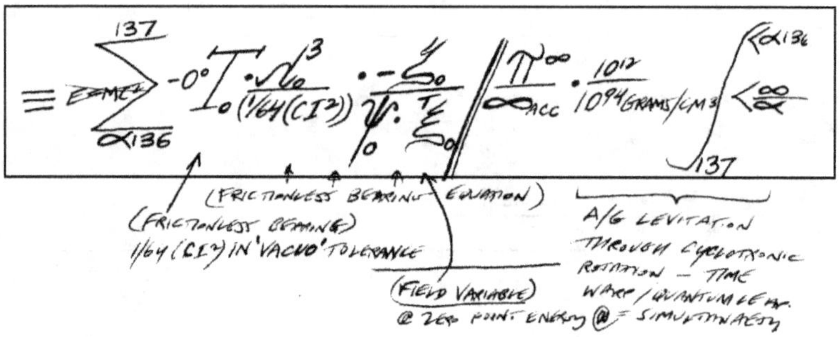
(FRICTIONLESS BEARING EQUATION)
(FRICTIONLESS BEARING)
1/64 (CI²) IN 'VACUO' TOLERANCE
(FIELD VARIABLE)
@ ZERO POINT ENERGY @ = SIMULTANEITY
A/G LEVITATION
THROUGH CYCLOTRONIC
ROTATION – TIME
WARP / QUANTUM LEAP.

18.5).

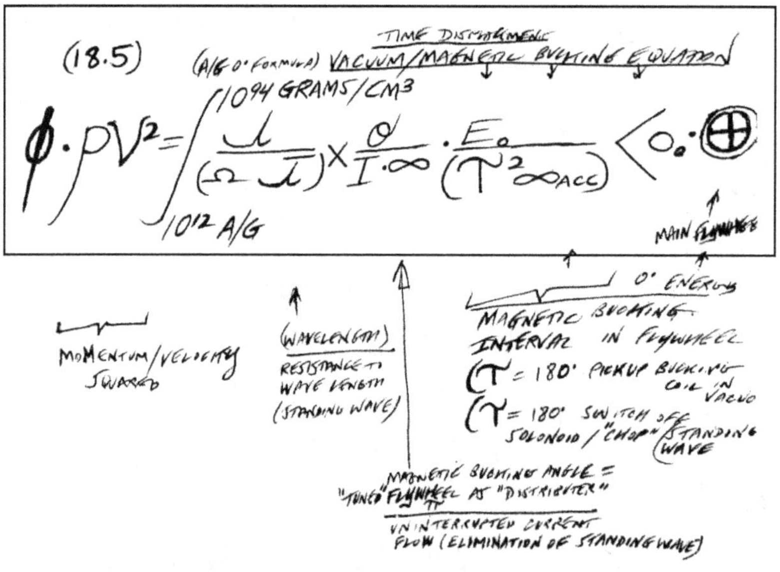
(18.5)
TIME DISPLACEMENT
(A/G 0° FORMULA) VACUUM/MAGNETIC BUCKING EQUATION
10⁹⁴ GRAMS/CM³
10¹² A/G
MAIN FLYWHEEL
0° ENERGY
MAGNETIC BUCKING
INTERVAL IN FLYWHEEL
MOMENTUM/VELOCITY
SQUARED
(WAVELENGTH)
RESISTANCE TO
WAVE LENGTH
(STANDING WAVE)
MAGNETIC BUCKING ANGLE =
"TUNED" FLYWHEEL AS "DISTRIBUTER"
UNINTERRUPTED CURRENT
FLOW (ELIMINATION OF STANDING WAVE)

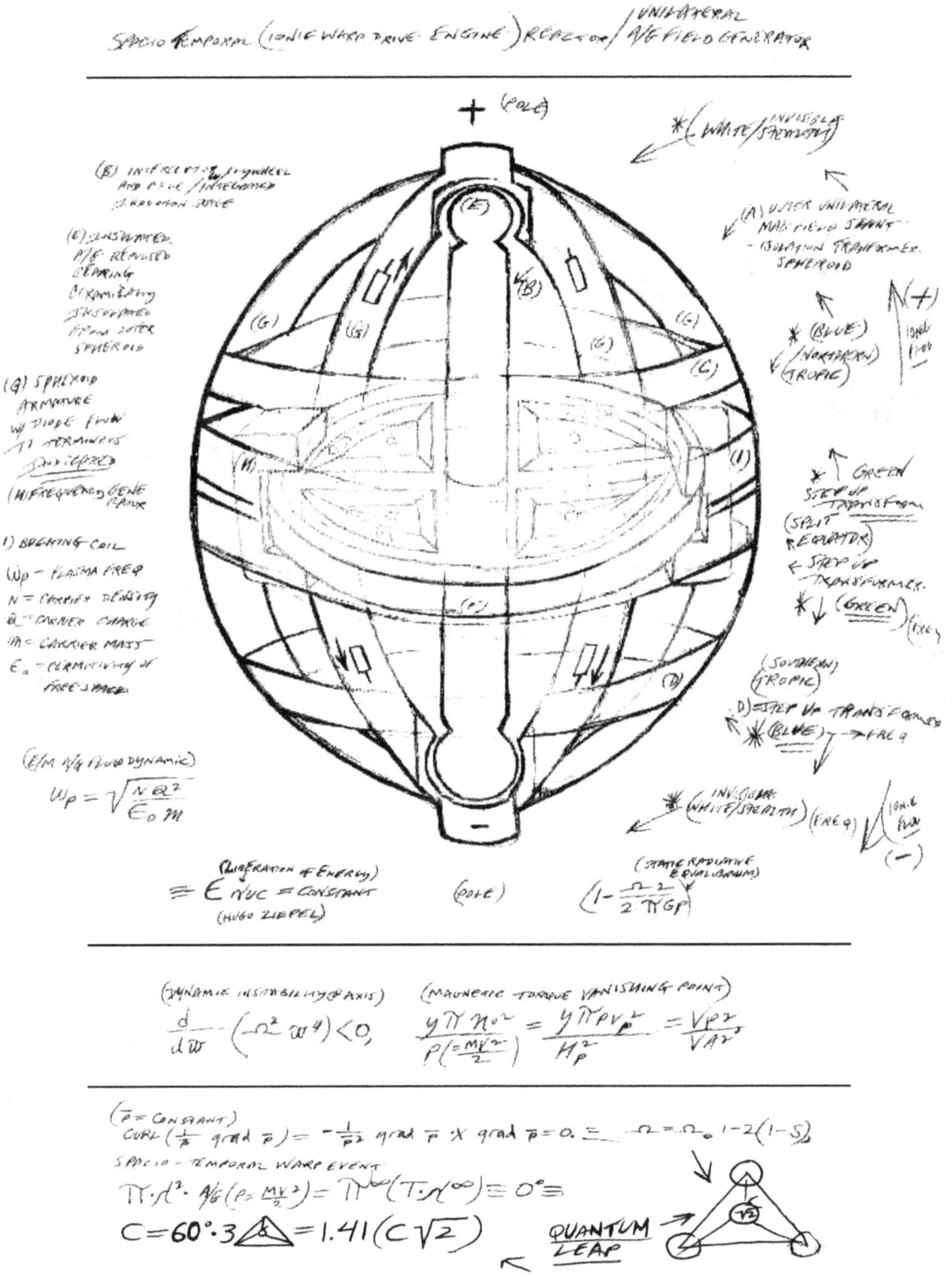
SPACIO TEMPORAL (IONIC WARP DRIVE ENGINE) REACTOR / UNILATERAL A/G FIELD GENERATOR
+ (POLE)
(A) OUTER UNILATERAL MAG FIELD SHUNT - ISOLATION TRANSFORMER SPHEROID
(G) SPHEROID ARMATURE
1) BUCKING COIL
Wp - PLASMA FREQ
N = CARRIER DENSITY
Q = CARRIER CHARGE
m = CARRIER MASS
E0 = PERMITIVITY OF FREE SPACE
(E/M A/G FLUID DYNAMIC)
(BLUE)
(NORTHERN TROPIC)
GREEN
STEP UP TRANSFORMER
(SPLIT EQUATOR)
(GREEN)
(SOUTHERN TROPIC)
D) STEP UP TRANSFORMER
(BLUE)
FREQ
(FREQ)
(POLE)
(HUGO ZIEPEL)
(STATIC RADIATIVE EQUILIBRIUM)
(DYNAMIC INSTABILITY @ AXIS)
(MAGNETIC TORQUE VANISHING POINT)
SPACIO - TEMPORAL WARP EVENT
C=60°·3 = 1.41 (C√2)
QUANTUM LEAP

Ω

Σ ∞

A

"SKY BELL":
ANTI-GRAVITATIONAL, HYDROGEN PLASMA, NUCLEAR FUSION, PERPETUAL ENGINE

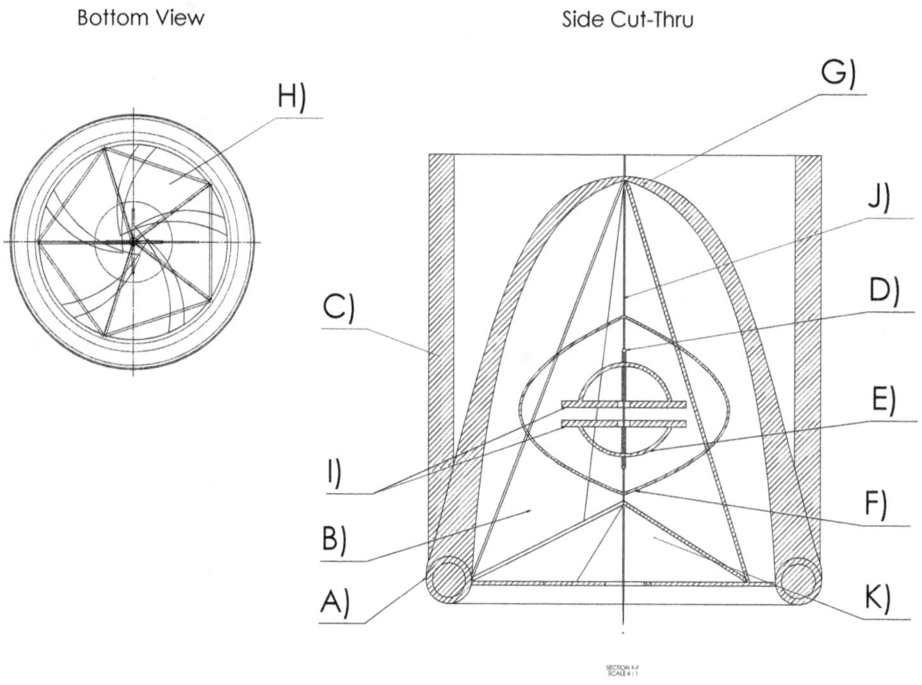

A) "Plasmatic Slush Hydrogen" Anti-Gravitational Toroid, "Floating Ring"(Superconductive Nuclear, "Chemical Plasma Flywheel")

B) Inverted "Jacob's Ladder" Protracted Compass, "Spline Sets", for "Triple Frequency" SONAR, 'Staggered Transmission' (UHF, VHF, "Linear Accelerated": Lazar Photon, Reactor, Toroid Bombardment: Triple Oscillation, Frequency Regulator) (Used as Circulatory/Vibrational Pump Mechanism, Frequency Agitating, Fusion Reactor, Dark Energy, Induction Vortex, for "German Bell Type, Hydrogen Plasma, Chemical Toroid, Nuclear Fusion, Superconductive Chamber") -

C) External "Double Hull" Reactor Shell, Domed Bell Cylindrical Encapsulation, W/Gausing "Magnetic Bottle Shielding"

D) "Upper Central Axis"

E) "Lower Center Axis"

F) "Dark Energy, External Space, "Zero Point Energy", Induction Rod"

G) "Upper Domed Bell" Tapered Reservoir, Uranium U235, Deuterium: Heavy Water, (2H, D)

H) "Sky Iris", W/Gause Shielded, Adjustable "Space Portal" External Environmental Space Aperture

I) "Magnetic Isolation Transformer", Twin Polarized, Electro-Magnetic Discs

J) "Main Axis", Adjustable Disc Transformer, "Frequency Fine Tuning" Central Devise, Oscillation Calibrator, Field Configuration, Intersecting Support Armature Array

K) "Jacob's Ladder" Array, Intersecting Support, "Base Elements"

1) Pure Outer Space, Environmental Unconditioned, As A Hermetically Sealed Vacuum - Temperature Regulated @ Absolute Zero Degrees: At The Theoretically/Physically Impossible-To-Reach Temperature, Of **"Zero Kelvin"**, Or Minus, **459.67 Degrees Fahrenheit** (Minus, **273.15 Degrees Celsius**), Atoms Would Potentially, "Stop Moving": As Such, "**Nothing Can Be Colder"**, Than **Absolute Zero**, On The **Kelvin Scale.** Therefore, It a Perfect, **Superconductive Medium**, For **"Dark Matter"** A Synthesized Experimental Condition, Empirically Reproduced, Will Yield Zero Point Energy, On An Infinite Quantum Level - And Consequently, Is The Only Gravity Repulsing System, That Is Sufficiently Powerful Enough, To Achieve Light Warp Speed, For Feasible Intergalactic Space Travel - Through Quantum Leaping/Via, White And Black Holes - Creating Specific, High Energy, Interdimensional, Atemporal, Navigational Pathways: Simultaneity Viaducts, Through The Time Space Continuum @ Beyond Light Speed

2) White Hole, Black Hole, Zero Point Energy Production, Occurs By Super Compression, Of Matter, By Solar Scale Magnitude, Gravitational Implosion @ At Low, To Virtually, No RPM's - Or By Hyper-Acceleration, Of Uncompressed Matter: Increasing Mass, Infinitely, To **(10, To The 94th Power@Grams/Per, CM Cubed):** Creating, Black Hole Specific, Anti-Gravity Phenomena - Deacceleration, Causes An Object, To Reemerge, Into Three Dimensional, Spacio-Temporally Conditioned, 3D Space, Through A White Hole - Zero Point Energy, Can Also Be Induced, By Hybriding Of The Two Physical Conditions: Deacceleration, Can Also Be Achieved, By Compromising, Or Breaching, **The Absolute Zero: Anti-Thermal, Hermetic Vacuum State:** And/Or Increasing, The Environmental Temperature - Therefore Inducing, Thermodynamic Entropic Breakdown, Into Sublight, Three Dimensional, Quantum Mechanics Range

3) Zero Point Energy, Is Incoherent, And Non-Sensible/Ungaugeable, Omnipresent, In An, Infinitely Expanding Quantum Field = The Universe, Comprising Infinity, Within Eternity, Moving At Increasing Expansion Rates, To Quantum Leap, Over Time -

4) Detection: Measuring Zero Point Energy/ Frequency Generation, Is Difficult: Due To Uniform General Field Dynamics - With Perceptible Differentiation, Only Occurring, In Spacio-Temporal, Frequency Shift Anomalies (Doppler Shift) Existing In Unique, Quantum Conditioned, Space Phenomena (Discreet Anomalous Quantum Events)

5) Orthogonic = Virtual Interdimensional, Orthogonic Flow/Flux, Of The Discreet Quantum, Indicates A White/Black Hole Reflux, Third And Fourth, Quantum Dynamic/ Energetic/ Synergistic Interface = Creating, A Perpetual Breeder Reactor, For Matter/ Anti-Matter, In An Infinitely Expanding Universe, At The Accelerating Speed Of Light -

6) Zero Point Energy, Exists, In A Free Radicalized Form, In A Ubiquitous, Hyper Agitated, Excitatious, Interdimensional, Perpetual Flux Capacitance State

7) The Zero Point Energy Reflux, (White Hole/Black Hole) Which Exist At An, Ultra High Frequency Level, Does NOT Readily Interact, With Three Dimensionally, Conditioned Matter: Until It Is Stepped Down Sufficiently, To Sublight, Band Width, And Quantum Range - Effectively, Depotentiating, Incipient Energy, Ergo = "Chopping" It's Potentiated, Full Band Width - And Inherent, Inductable, Power Factor Potential

8) Physical Matter, Must Be Hyper-Accelerated, To Interact, With Zero Point Energy (To Achieve Dark Matter, Induction Point @ The Highest Coefficient) This Can Be Achieved Through, Propulsion - Rotation - This Can Also Be Achieved By High Energy Ionic Flow, Through Superconductive Plasma, Field Mediums @ Light Speed Power Factor: Amplified By Truncated Arrays, Of Focusing Coils - And/Or, Through, Aforesaid

Super Conductive, Plasma Fields = Depotentiated, Thermodynamic Entropic Breakdown = By Sustaining The Reaction, In Super/Ultra Cooled, Absolute Zero Hermetically, Vacuum Sealed Reaction Chamber / And/Or, Focused Particle Beam Acceleration/ Generated And Amplified, Through Geodesically Architected, Resonant Cavities - Aforesaid, Lazar Apparatus, Synergistically Acting Upon, Incipient Matter Further Amplified, By The Geodesic Reactor Matrix, Will Effectively Produce, Zero Point Energy, Within A Previously, Three Dimensional, Contained Magnetic Field

9) **Law Of Zero Point Energy =** High Frequency Zero Point Energy, Does Not Normally React With, Three Dimensional Matter = Unless The Matter In Question, Is Brought Up, To, Ultra High Frequency, Band Width - Thus Focusing, And Inducting, Aforesaid Energy (Dark Matter) By Exponentially Increasing, The Acceleration, **To Light Warp Speed -** Matter, Will Quantum Leap, **From 134,000 MPS, To 186,000 MPS, In Simultaneity**

10) Conical, (Tesla Focusing Coils) in A Steam Vortex, Will Create, **"Ball Lightning"** Inducting A Scalar Effect @ Sea Level, Accomplishing, A Form Of Dark Matter Induction The Earth Itself, Functioning, As An "Open Delta Circuit" 33 Degrees, Rocking Toroid Empirically Establishing, A Macrocosmic, Solar/Planetoidal, Quantum Model: To Empirically Replicate, And Energetically Harness, Zero Point Productively -

11) A Caduceus Coil, Manufactured, With Perfect Counterwindings = Equilaterally Opposing, Neutralized Windings, Over A Ferrite Core, Hyperactive Potentiated Energy, To Zero Point, In A Chambered, Vacuum Condition @ Absolute Negative Degrees, Celsius, @ Zero Kelvin: Current Scientific Technology, Withstanding

12) Polarized Vacuum/ With Internal Magnetic Bucking, (High Frequency Generation/ Modulation) = Aether/Dark Matter, Through Induction Toroid, Caducean Coil -

13) "Flywheel Induced" Zero Point Energy - High Frequency Magnetic Bucking Fields Within Hermetically Sealed, Vacuum Chambers @ Negative Zero Degrees - Equipped With, Magnetically Sustained, Axis Point, Frictionless, Anti-Gravitational, Bearing Sets

14) Creating Specific, Anti-Gravitational/Levitational = Electromagnetic, Dark Matter, Induction Field = Empirically Recreating: Solar, Planetoidal, Geo, Astrophysical = Micro-Replication, Of Universal/Unified Field Theory, Of Fusion Physics

15) Infinite Acceleration, Of Energy Inducted Matter, Into Dark Energy, By High Energy, Permitivity Waves - Pulling The Object Towards, The Interdimensional Components, Of Profound, Electromagnetism State = Cyclotron = Interceptor = Perpetual Engine/ Generator/Nuclear, Fusion Breeder Reactor (=Sun) = Warp Drive, Propulsion System = Time Displacement Device = High Energy, Caduceus Induction, H Plasma, Toroid Coil

16) Non-linear, Hydro-Dynamic, And/Or, Electrodynamic Paradigm Coherence, Established, From A Random Turbulence = Forming, A Plasmoid, Virtual Coherency = Synergistically Valence Balanced, @ Micro/Macro, Substructure = Reconfigured Out Of, Entropic, Thermodynamic Breakdown, And Reordering Of The Thermodynamic Sublight Chaos - In Reintroduced Stasis, Refuting, Long Term, Chaos Theory Pardigms

17) Lorentz Constant, Indicates, Matter Resonates, with High Acceleration, Amplifying Density, To Infinity = Matter Transformates, Into Antimatter = Inducing Light Warping And Anti-Gravitational, Magnetic, Motive Force = In The Time Displacement Field

18) The Second Law, Of Thermodynamics, Was Modified, By Ilya Prigonine (Nobel Laureate, 1977, Chemistry) = Revision/Refutation, Of "Chaos Theory" = All Matter/ Energy, Eventually, Reorganizes Itself, After Thermodynamic Entropic Breakdown =

Reestablishing/Reorganizing, In A Neutralized, High Frequency, Interdimensional, Quantum Space = As Said, Eventually, Reestablishing, Coherent Order/Beyond Thermodynamic Breakdown/ Plasmoidal Vortex/ Progressional Rotation/ (A Hellical Rotation, Around, A Ferrite Cylinder, Or a Plasmoidal E/M Axis = Ferrite, that Truncates, Into, A Magnetized Toroid, Is The Optimal Design Paradigm = The Physics Formula Itself, Loosely Follows The Path, Of This Verbal Explanation, Of Zero Point Energy/ Nuclear Cold Fusion Reaction - Planets Axis Revolve, Around Ferrite Molten Cores

19) Tachyon Field, Quantum Conditioned, String Flux, Special Event = Anticipates, "Black Hole Horizons" = As Tachyon Particles, Decelerate, Into Sublight, "Virtual Materiality" From Out Of The Fourth, To The Third Dimension State: Breaching Hyperspace, in A Plasmoid Energy Tear = Analogous, To The Theoretical Models, Of The, Big Bang Theory - Converting Momentarily, Anti-Matter, Into Matter = Condensing Incipient Localized Space = Inducted Zero Point Energy, Into Anti-Gravitational, Focal Point, Of Dark Matter Conflux - Creating Subsequent, Quantum And Spacio-Temporal Displacement, Proximal Non-Linear Anomalies - Which Can Technologically Synthesized, Harnessed And Exploited - To Control And Navigate, the Space/Time Continuum - Teleportationally, In Four Dimensionally Conditioned, Quantum Leap: This Occurring Between, The Nearest Twin Matched Pair, Of White And Black Hole, Wormhole Quantum Axis Formations = Empirically Producing, **Simultaneity @ Axis C:** Thus Replicating Solar And Planetary, Astrophysical Events

20) Nuclear Fusion Physics, Replicates Solar Astrophysics, On A Macrocosmic Level Producing A Gravitational Field, Which When Hyper-Energized: Will Fold Into It's Own Discreet, Relative Time Space = Into Tachyon Productive, Four Dimensional Anti-Matter Matrix Sustained, In Simultaneity, As A White/Black Hole, Quantum Event -

21) Hydrogen In The Rare Organic Compound, Molecular Form, Or Deuterium, Is A Basic Isotopic Fuel, Of Cold Fusion Nuclear Physics: When Atomically Fused With Plutonium, While Being Struck/Reaction Ignited/Triggered, By A Linear Accelerated/ Cyclotron Actuated, Particle Of Radioactive, Tritium Isotope - Will Ignite A Nuclear Fire, @ Least, Twelve Magnitudes Hotter, Than The Internal Core, Of The Sun = Cold Fusion Reaction, As A Power Source, Of Temporal Displacement, Technology: Is Absolutely Critical, In Producing, A Controlled And Sustained, Black Hole/White Hole, Zero Point Induction Event -

22) Deuterium, A Rare, Naturally Occurring, Radical Hydrogen Isotope (Heavy Water) Can Be Found And Farmed, From The Still Waters, Of Large Swamps, And Deep Fyords = Deuterium, Can Also Be Synthesized, By Employing Breeder Reactors, Attached, To Cold Fusion Reactor, Slush Hydrogen Toroid Plasma Core; The Slush Hydrogen, Toroid Superconductor, Magnetic Cyclotron Ring, @ The Center Of The Cold Fusion, Nuclear Pile - The Toroid Ring, Filled With Radioactive, Hydrogen Slush, Can Be Tapped And Pumped, Out Of The Central Toroid Ring, Of The Fusion Reactor Core And Then Allowed, To Heat Up, From The En Vacuo, Reactor State, Of Absolute Zero Drawn From Within, The Reactor Core: Which Simulates, The Conditions Of Deep Space - Once Removed, From The Reactor Core, The Hydrogen Slush, Returns To A Gaseous, Radioactive Vapor State = And Spontaneously, Begins To Release Radioactivated Particles, Into A Warmed, External Environment - Not Unlike, The Condensation Process, In Alcohol Distillation = Said Nuclear Fallout, Can Be Trapped, And Targeted, Into A Pressurized Tank (Simulating The Conditions, Of A Deep Fyord)

Designed, Like A Natural Gas, Storage/Distribution, Tower - Infused, Into A Pressurized, Cooled Tank, Of Distilled Water - Like The Gas Storage Tower, The Tank Equipped, With Ring Attached, Symmetrical Array, Of Hydraulic Presses, Exerting Tremendous Downward Pressure, Upon The Distilled Water, Linear Accelerator Bombarded, By Infused Radioactive Hydrogen = Computer Monitored, To Hydraulically Pressure Sustain, A Static/Stilled, Pressurized Equilibrium, Of A Thousand Feet, Equivalent Atmospheres, Of Standing Water @ A Deep River Bed, Terminus = The Pressurized And Free Radical Hydrogen, Irradiated/Bombarded Water, Is Further Refrigeration Chilled, To Just Above Freezing (While Under Hydraulic Pressure) To Replicate Synthetically, The Underwater, Breeder Reactor Medium, Deep River Conditions Of Super Cold, And Stilled Water, That Exists, At The Bottom Of Fyords - The Bombardment, Of The Said, Stilled Water, By The Hydrogen Gas, Molecular Stream - Of Vaporizing, Radioactive, Radical Hydrogen Isotopes, From Slush Toroid Depleted Deuterium, Can Be Reintroduced, Back Into The Slush Toroid, Reactor Ring: Once Reinfused, Into Pressurized Distilled Water: As "Synthetic", Heavy Water, Deuterium To Create A Perpetual, Energetic, Closed System, Nuclear Cycle = Similar To The Birth And Death, Of Stars = Recycled, Nuclear Slush Fallout, Providing and Endless Supply, Of, Synthesized Deuterium, For Perpetual Cold Fusion Reaction (Fusion Star Physics)

23) The Still Waters, Of Both Deep River, Cold Fyords, And Inland Large Swamps - Both Unaffected, By Coastal, Tidal Action: Continuously Bombarded, By Low Level, Photonic Nuclear Radiation, Of The Sun = Accomplishes, The Same Task, To Produce Heavy Water, Deuterium - Over The Slow, Exposure Course, Of Thousand Of Years = The Still Water, Heavy Hydrogen Particles, Never Changing Physical Position - Sun Irradiated, Over Enormous Periods Of Time, Will Eventually Bond And Mutate, Into Radioactive Isotopes - Photosynthesized, Insitu, By Stilled Deep Water, Penetrating Radiation, Of The Sun, Over Countless Millennium = In Contrast, A Fusion Reactor, Essentially, A Manmade Version, Of A Fusion Hydrogen Star - Can Create, The Selfsame, Particle Mutation Reaction, Within, It's Own Closed System: Nuclear Fallout Upon Pressurized And Chilled, Still Water, Instantaneously, Under Controlled Laboratory Conditions = Providing A Breeder Reactor Stage, For The Fusion, Tokamak Reactor

24) Another Potential Method, Of Producing Heavy Water/Deuterium, Is To Bombard, A Super-Pressurized, Stasis Water Tank, Of Pure H20, Pressurized, And Refrigerated - To One+ Hundred Atmospheres, With Semi-Frozen, Slush Particles, Of Free Hydrogen Isotope, Shot Through, A 42 Inch, Linear Accelerator @ Speed Of Light, At A Guassed Bombardment, Cyclotron Chambered/Particle Collider - Traveling, At Enormous Speed, The Accelerated Particles, In Hyperspacial, Virtual Flux , May Attach Themselves, Electromagnetically, To Receptive/Available, Heavy Water Suitable, H20, Molecule Targets Not Unlike, Arc Welding, Or Annealing Metal Alloys Together, Using Amplified Electricity, And/Or, Extreme Heat, To Bond Otherwise, Antagonistic/Repulsing Matter

25) The Optimal Reactor Configuration, Should Be Faceted, Along Euclidean Vertices In The Shape, Of A Jeweled Crystal, Allowing Linear Acceleration Bombardment, Of Radioactive Tritium Isotope, To Be Hyper-Accelerated, In Vacuo, To A Central Target Area, In The Precise Center, Of The Fusion Reactor Core - Strategically Oriented, At Each Facia, Of The Fusion Reactor, Euclidian Planes - There Is An Optimal, Co-Efficient, Of Accurate Bombardment, Strike Potential, To The Plutonium/Deuterium, Nuclear Core Fuel, The Tritium Accelerated @ The Speed Of Light, "Welds" The Three

Separate Particles, Creating A Synthetic, "Solar Flare", Capable, Of Yield Producing, Exothermic Heat Discharge, In Excess, Of Twelve Magnitudes, The Core Temperature, Of An Equivalent Sized, Solar Mass, Of Our Own Sun = Additionally, The Linear Particle Accelerators, Provide, A Symmetrically Displaced, Magnetic Field "Bottle" Sufficient, To Contain, Not Only The Super Nova, Heat Levels, But Also Magnetically Trap, The Gamma Radiation, Off Gassing, Caused By The Fusion Nuclear Reaction - The Reactor, Itself Must Be, "Double Hull", Constructed, As The Trapped, Gamma Radiation, Chemically Introduced, Serves, As A Fusion Reactor, "Super-Coolant", As It Is, Reservoir Accumulated, Between, The Twin Reactor Walls, During Production/ Operation, Of The Nuclear Fusion, "Tokamak" Reactor, Refrigerating The Containment

25) Lagrangian Mechanics/Dynamics, Must Be Maintained, In The Magnetic Ring Which Is Formed, By The Linear Accelerator Array, Around, A Nuclear Fusion Reactor, Not Unlike A Gravitational Field, Consistent With Rotating Stars - Any Detectable Breach, In The Field Uniformity, Of The Magnetic Ring, Will Cause Fusion Reaction Entropy, To Thermodynamic Breakdown: The Generated Heat, No Longer Repulsed Back, Upon Itself, En Vacuo, @ Absolute Zero - Will Seek Any Temperature Variant, In The Reactor Core, And Flare Through It, Seeking To Form, A Plasmoid, With Any Adjacent, Three Dimensional Material, In The Path, Of Least Resistance, Not Unlike Lightning, During Atmospheric, Scalar Effect, Seeking The Tallest Object Available, As A Electronic Conductor, To Discharge Itself, Circuiting, Into Natural Earth Ground -

26) Degaussing, Engineering Protocols, Must Be In Place, To Contain, The Degree Amount, Of Magnetic Reactor Shielding, From Effecting, Outer Environmental Conditions: Requiring A Secondary, Encapsulating Reactor, Gaussing Hull, Such As A, Faraday EMP Cage, Calibrated To Withstand, Magnetic Fallout, This Will Additionally, Focus And Compress, The "Magnetic Bubble" Around, The Fusion Tokamak, Nuclear Reactor, For Highest Co-Valence, Of Inner And Outer Core, Optimal Efficiency: In Isometric, Dueled Counterforce, Energetic Repulsion, Fusion Reaction/Magnetic Bottle, In Zero Point, Dynamic Gaussian Tensioning -

28) Twin Matched Pairs, Of Side Mounted Fusion Reactors, Placed, Upon A Line Symmetrical Interface, Become Larger, More Powerful Versions, Of Linear Accelerators, Exhibiting, The Same Capacity, To Accelerate Matter, Through Their Reactor Cores, At And Beyond, The Speed Of Light - This Phenomena, Is Directly Correlated, To Co-Valence, Amount Of Energy, That Is Generated, In The Fusion Reaction Chamber

29) When The Amount Of Nuclear Discharge, Carries A Higher Valence Output, That The Surrounding, Three Dimensionally Conditioned/Bonded Matter, In It's General Proxemics: It Will Interdimensionally, Parallax, Seek Out, A Black Hole, White Hole, Point, Of Nuclear Discharge: Effectively Becoming, A Teleportation Devise - If The Tokamaks, In Question, Are Not Precisely Calibrated/Twinned, To One Another, A Black Hole, Can Be Opened, In The Nearest Alternative, Adjoining Space - The Natural Space Targeted, Will Invariably Be, In The Exospheric Range, Of Outer Space, Beyond Earth Atmosphere, Due To The Thermodynamic Requirements, Of An Absolute Zero Point, Temperature State: Replicating, The Conditions, Of The Reactor Core: Additionally This May Cause, Spacio-Temporal Anomalies, Due To The Presence Of Faster, Than Light Particles, Effecting The Surrounding Matter - This Seems To Be, A Shock Wave Effect, As The Particles Involved, Decelerating Only Momentarily, And Are Highly Unstable, In Sublight, Conditioned Space - The Time Anomalies Created, Are Therefore, Of A Shock

Wave Nature, With Smaller Aftershocks, Not Unlike, An Earthquake, Or Tidal Wave, Hitting Landfall - The Presence, Of These Variety, Of Hyperaccelerated, Energy Particles: Anti-Matter, Dark Matter, Plasma Superconductors, Etc, If Not Trapped Magnetically: Just Like A Nuclear Fusion, Reactor Bottle, Will Eventually Return, Into The "Black Hole Tare" Of Hyperspace, Once The Transmitting, Energy Discharge, Of The Fusion Reactor Tokamak, Is Eliminated, On The Opposing Side, Of The Reaction - Collapsing Back, Into Normalized Space - Measurable, Spacio-Temporal Anomalies, Will Have, A "Phantom Effect" Eventually Resynchronizing, After A Calculable, Observed Duration, Analogous, To Gravitational Descent Rates, Of Newtonian Acceleration Tables, Yet In Diametric Opposite/Reverse: As It Essentially, An Anti-Gravitational Event, First And Foremost - A Physics, Existing In Mirrored Opposite - Event Horizon Equations, Can Accurately Compute, The Epicenter Of These Four Dimensional, Portal Events, Determining Intervals, Of Re-Synchronization - Post Reaction, By Doppler Shift Recordation, Of Environmental Light Phasing, In The Effected, Time Space Zone Parameter - Determining Relativity, To Teleported Energy Input, And Time Reaction Spacial Effect, @ Rate Of Deterioration - Various Hyperspace Particle/Frequencies, When Traveling Together, At Various Decay Rates, Also Have The Reverse Effect, Of Reaccelerating, Acting, As Synergistic Catalysts, Of Hyper-Propulsion: Once They Are Energetically Disentangled, By The Influence, Of An Artificial Tethering Force, In The Form, Of An Overproductive, "De-Twinned" Fusion Tokamak Reactor - Experimental Precaution, Is Advisable, In Avoiding Proximity, To Oceanic Trenches - **(*As They Create Profound Hydrodynamic, Thermal, Extreme Pressure, Super Electoconductive, Plasma Conditions, Analogous, To The Energetic Vacuum, Of Zero Point, Exospheric Space - Potentially Creating, A "Devil's Triangle" Effect, Upon Opportunistic Matter, Randomly Targeted, For Induction/Teleportation - Without A Twin, Exit Portal, For Predictable Re-Materialization, Proximal Interface)** With Geomagnetic Forces Amplified, Due To The Four, To Five Mile Depth, Of These Underwater "Black Hole" Induction Ports, In Relation, To The Earths Core: It Is The Therefore Advisable, To Avoid Them, With Fusion Reactor Technology -
30) Living Biological Entities, Cannot Survive Teleportation, Unless A Magnetic Bottle, Containment Field, With Faraday Array Shielding, And Interior Life Support, Is Provided For - The Proximity, Of Teleportation Range, Between Tokamaks, Will Condition, The Length Of Experienced Duration, Of The Teleportees, "Experience"
31) Internal Atomic Clock, Synchronization, With An Outer Environmental Empirical Referent, Will Determine The Degree, Of Spacio-Temporal Displacement - Theoretically, There Should Be, No Time Elapse, In The Teleportation Vehicle - As It Is Traveling Faster, Than The Light Barrier: Reverse Time Occurrence, Is Also A Potential Probability Negative Time Elapse, Can Be Determined, By Using Biological Species, With Extremely Short Life Spans, To See If They Physically Regress, In Observable Physical Age (Insects, Pregnant Laboratory Mice/Offspring) - Fast Decaying Isotopes, With Short Half Lives, Can Also Be Used, As Negative Time Influence, Upon Organic Substances - The Teleportational, Targeting Range, Can Be Incrementally Lengthened, To Establish Graduated Time Distortions, Within And Without, The Magnetized/Gaussed, Test Probe Vehicle - To Calculate Accurate Rates, Of Warp Acceleration, To Instantaneous Simultaneity States - Improper Rates, Of Acceleration, With Insufficient Propulsion, Or Calculated Energy Input, Per Teleported, Organic/Material Mass, Will Cause Chemical,

Material, Biological Melding, And Genetic Mutation, Obviously, A Serious Risk Factor, In The Development Of These Devices - Therefore, Accurate Physics Tables, Derived From, Unmanned Probe Testing, Are Absolutely Imperative, In Alpha Stage Trials - Any Biology, Should Be Analyzed, For Introduced Pathogens, Or Antigens, In The Post Mortem Autopsy, Medical Follow Up -

35) Theta Wave Attunement, Of Ground Based, Or Geosynchronous Orbiting Satellites, Equipped With Anti-Gravitation, Fusion Nuclear "Flywheel" Propulsion Platforms, Will Harmonically Resonate, With One Another, Through Space State Vacuum, Or Atmospheric Weather Conditions, Like Radio Transmitters, And Receivers, Having A Synergistic, Stabilizing Resonant Cavity Effect, Exerted Upon Each Other - Additionally, They Become Earth Penetrating, Terrestrial Tuning Forks, And Serve To True, The Gravitational Rotational Orbit, Of The Earth, @ 1,000 MPH, Preventing Tectonic, "Bucking" And Dampening, The Moment Of Axis Drift, Intermittent Rotational Stall/ Wobble, And Moment, Of Geological Sheer, Associated With Earthquake, Volcanic Eruption, And Tidal Wave Activity: Mostly Experienced, Around Active, Major World Geological, Fault Lines, Which Can Be Monitored, And Specifically Targeted - Weather Control, May Also Be, An Additional Benefit, As The Technology Advances, And Sophisticates - Theta Wave, Is Also Associated, With High Meditation States, In Humans, And Other Mammals: And May Substantially Enhance Awareness, And Intelligence Quotients, Positively Effecting Mental Health, In The General Population

36) A Nuclear Fusion Power Flywheel/Toroid Plasma Ring, Has Inherent Anti-Gravitational Properties, Separate And Distinct, From General Aerodynamics, Or Turbine Jet, Or Rocketry Propulsion - Either By Mechanical Rotational, In Accordance With Lorentz Invariant, The Hyper Spun Object, Will Exponentially Increase, It's Gravitational Mass, To The Outer Limit, Of Infinite Mass/@Infinite Acceleration, The Objects Own, Gravitational/Geo-Magnetic Pull, To The Earth, Is Progressively Relieved: As It Forms It's Own, Autonomous/Independent, Gravitational Field - At The Discreet Point, Of Weightlessness, In Relativistic Terms, To Effectively Neutralize, Earths Gravitational Pull, Upon It: When, The Object Spin, Is Sufficient, To Create Upward Lift, And Thrust, Independent Of Surrounding, Environmental Physics, Effectively Relieving, It's Own Gravity - Theoretically, This Independent Gravitational Field, Can Be Energetically Transferred, To Any Tethered Object, Providing Greatly Enhanced Payload Capacity, Based Upon, The Hyperspun, Rate Of Rotation, Of The Propelling, Tokamak Power Source: As, In The Example, Of A Cyclonic Event, Where Houses, Uprooted Trees, Automobiles, Are Air Syphoned Up, And Strewn Miles Away, From A Cyclone Strike - Or In The Case, Of A Powerful, Oceanic Whirlpool, That Can Swallow Large Ships, Or Spontaneously Form, As A Large Ship, Is Sinking: Like A Massive Drainage Hole/Inverse Syphon/Vacuum Sump, Attenuated By It's Own Descending Mass, Into A Resistive Plasma Counterforce: A Nuclear Fusion Reactor, Actuating, A Mechanical, Or Chemical Superconductor, Plasma Ring/Toroid Flywheel, Of Substantial Mass, Has Nearly Infinite, Lifting And Acceleration, Multi-Directional, Propulsion Characteristics - When Placed, In An Open Delta Circuit, Triangulated Array, With, A Centralized Induction Port, Located In The Center, It Is Able To Maneuver, In Any Direction - Drawing, Additional Free Energy, Via, Inducted Magnetism, From The Localized Environment - Terminal Velocity, In Atmosphere, Approximates **42,000 MPH**, Or, **Mach 55**, @ **(767.269 MPH = Mach1)** = Speed Of Sound - Air Turbulence, And Wind Sheer

Friction Resistance, Minimized, By A Magnetic Bubble, Produced By The Flywheels, Which Double, As Linear Accelerators/Generators, Which Magnetically "Encases" The Hull, Pushing/Repulsing, Surrounding Atmospheric Air, Away From The Forward Course, Of Linear Trajectory - In The Vacuum Of Space, Where No Aerodynamic Physics Exist: The Acceleration, Of The Vehicle, Increases Exponentially, As The Fusion Reactor Flywheels, Constantly Spin, Like Perpetual Engines: Not Requiring Rocket Fuel, Or Combustable Propellant, Thus Allowing, Said Vehicle, Over Time And Distance, To Break, The Light Barrier, When Arriving, At **134,000 Miles Per Second**: The Additional Power Factor Induction, Of Available Free, Zero Point Energy, Which Causes, A Quantum Leap, To Occur, In Simultaneity, Accelerates Beyond, **186,000+ MPS** = Speed Of Light Barrier @ Approximately, **26%, Percent Induction Rate**, Power/ Quantum, Factor: This Synergistic Effect, Is Created, When The Atmospheric/Space, Magnetism, Is Attracted To, And Amalgamated Into, The "Open Delta" Configured Magnetic Propulsion System, Of The Time Warping Vehicle, During Nuclear Fusion Tokamak Operation: Teasing Out Tachyon Field, Black Hole Production, In The Induction Stage, Through A Variety, Of Effective/Productive, Commutating Arrays

36) This Occurs Most Efficiently, When The Three Sets, Of Generating Flywheels Create Sufficient Induction Energy, To Create, A Stable Black Hole, In The Central Induction Port, In The Structural Center, Of The Triangulated Fuselage: Drawing Hyper-Accelerated, Energy Particles, Dark/Free Energy, Anti-Matter - An Additional Benefit, To A Magnetically Propelled, Anti-Gravitational Vehicle, Is That It Does Not Experience, External G Force, Tugging Or Thrust Torc, In Earth Atmosphere, As It Is **NOT,** Fighting Gravity, To Achieve Escape Velocity, = **25,020 MPH**, In Space Flight, Theoretically, **1G, Of Cabin Gravity**, Is Constantly Experienced/Maintained, At Any Point, Of Exterior Acceleration - Weightlessness, In Space Flight, Related Bone, And Muscle Tissue Loss, As Well As Gamma/Solar, Radiation Exposure, Outer Hull Bombardment - These Are Magnetically, Energy Shield Eliminated: Allowing, For Extended, Long Term Space Missions, Without Psychological, Or Physical Endangerment, Associated With, **Zero G Force**, Current Rocket Technology - As No Combustable Fuel Is Utilized, Initial Propellant Payload, Is Weight Eliminated, Massive Rocket Stages, Are Effectively Eliminated, With Attending Costs, And Environmental Impact, Which Has Been Often Speculated, To Cause Short Term, Negative Impact, Weather Changes - Chemically Seeding, The Jet Stream = **(110-250 MPH)**, And Lower Barrier Exosphere, With Tons Of Rocket Fuel Exhaust, Per Manned Rocket Mission, Thus Being Relegated, To X-Level, Experimental, Lunar Landing Trials, The Outer Distance, Of Thyicol Propelled Rocketry

37) Foucault's Gyroscope, As Well As Polarized Magnets, Defy Gravitational, General Dynamics, And Are Therefore, The Two Major Components, Of An Anti-Gravitational Device - Hyper-Acceleration Achieved, By Magnetized Bifurcated Array, Of Microwave Oscillation Activated, Bucking Coils, Which Serve As Intermittent, Electro-Magnets, Synchronized, To The Bucking Points, Of Spin Travel, During Flywheel, Gyroscopic Path Crossing, The Bucking/Repulsion, Action Of Said, Bucking Coils, Can Be Accelerated, To Induce Hyper-Rotation, Of The Gyroscope/Flywheel, This Must Be Done, In Vacuo @ Absolute Zero Temperature Conditions, The Flywheel, Must Be Magnetic "Floated" To Avoid, Physical Contact, With Frictionless, Magnetic Bearings, That Suspend, And Stabilize, The Array - Micro-Corrections, Requiring, Split Second Calculations, Must Be Employed, To Achieve Maximum, Spin Integrity, And Stability -

Similar, To A Steam Engine, **"Flyball"**, Attached To The Drive Shaft, Of A Steam Power Plant, To Determine RPM, And Also Serves, As A Mechanical, Limiting Governor, For The Apparatus - Never Reaching **180 Decrees**, Of Perfect Horizontality, Determined To Be Physically Impossible To Attain, Due To Thermodynamic Breakdown, Of Motive Rotational Efficiency, And The Downward Pulling Influence, Of Earth Gravity, Upon The Latitudinal Deflection, Tangent Of Offset, From Perfectly Maintained Horizontality -

38) In This New Instance, Gravitationally Neutralized Forces, Are Successfully Overcome, In The Vacuum/Zero Temperature/Magnetic, Microenvironment, Of The Fusion Nuclear Powered, Magnetically Suspended Flywheel - Attaining Levitational Lift, And Gyroscopic Propulsion, Without The Requirement, Of Propellers, Turbines, Or Rocket Firing Chambers: To Achieve Anti-Gravity/Repulsion Thrust, Directly Against Earths Gravitational, Atmospheric Forces - This Magnetically/Cyclonically Generated, Motive Force, Is Exponential Increased, In The Ideal Frictionless Environment, Of Open Frozen Vacuum Of Space = @ Perpetual Acceleration, Beyond Light Warp Speed

39) Cartesian Coordinates, In A Parallax Perspective, Upon A Zero Point, Event Horizon, While Entering, The Quantum Induction Field, Of Interdimensional Hyperspace, Causes Ambient Light, And Surrounding Matter, Particle Bending, Toward The Ultimate, Energy Threshold, Of A Black Hole/Worm Hole, Formation Phenomena The Venturi Like Event, Draws The Amalgamated Induction, Of The Surrounding Matter Into A, Dematerialized Pure Energy State: However General Field Unity, And Material Structural Integrity, Should Maintain, In The Hyperaccelerated State, Upon Reentry, Sublight Conditioned Space - The Quantum Warp, Itself, Should Be Symmetrical, In Nature, Under Optimal Conditions, The Four Dimensional Energy State, Obeying The Same Laws Of Physics, Will Keep Individuated, Energy Signatures, Separated, Based Upon, Their Pre-Existing Material Nature: Translating, "Form Integrity", When Warp Travel, Is Experienced - This Is Exemplified, By Particle Integrity, Decay, And Reacceleration Paradigms, Observed In, And Around, Synthetically Created Black Hole Phenomena - The Converse Condition, Should Exist, On The Other Side Of The Interdimensional Interface, Empirically Based, On This Hyperaccelerated Particle Activity - Needless To Say, The Black Hole Condition, Must Be Energetically Stable, And Of Sufficient Configuration, To Allow, A Large Space Vehicle To Pass, This Should Be, A Self Regulating Homeostatic Event, Based On The Forward Thrust, Acceleration, And Navigational Attitude, Of The Projectile, Being Acted Upon, By General Field Physics, Within The Warp Breaking, Spacio-Temperal Displacement Event - The Internal And External, Physical Fields, Must Remain, Di-Opposed, In Proportion, And Energy Yield Valence, Precisely Scaled To The Dematerialized Objects, Amalgam Mass/Energy, Hyperaccelerated, "Warp Break Point", Conversion State -

40) The Object In Question, Converted, Into Hyperaccelerated Anti-Matter, Will Follow A Linear Path, Of Acceleration Trajectory, In Forward Momentum, Navigational Line Of Sight, Following The Path, Of Least Resistance, And Highest Co-Efficient, Of Linear Propulsion, Utilizing The Destination Target, Through The Path Of Least Physical Distance: Conforming, To Conservation Of Energy, Related Scientific Principles, Pre-Alignment, Is Therefore Required, As Adjusted Navigation, In Hyperaccelerated Exciter Quantum State, In Not Possible, Beyond A Forward Linear, Trajectory Course -

41) "Interdimensional Plane Interface", And The Quantum Field Induced, Is A Zero Point Euclidean Axis, Quantum Integer, That Is "Calculable To Pi", This "Nexus Quadrant",

Can Therefore, Be Postulated, To Pre-Exist Ubiquitously, In General Physical States: In Black And White Hole Polarity, Throughout The Known Universe, The Matched Valence Correspondence, Of Wormhole Sets, Quantify Each Others, The "Missing Pi Factor", Incalculable Integer, Only Discernible, By Geometrical Paradigm, Of Linked Langrangian Spheroids, Interfacing, At A Mutually Tethered, **OPEN** Quantum Nexus -
42) The Most Stable Propulsion Method, For Quantum Leap, Space Time Travel, Is To Establish, Twin Synthetic Stargates, Energized By Tokamak Fusion Reactors, To Create, Valence Matched Set, Plasma Toroid Rings: Slush Hydrogen, Serves As, A Superconductive, Cyclonic Plasmoid Medium, For Rotationally, Hyperaccelerated Nuclear Particles, Which Create, An Electro-Chemical Flywheel, With Enormous Rotational/Ant-Gravitational, Dynamics - Within The Toroid Ring, Star Gate Sets - The Plasma Toroid Rings, Can Be Transported And Positioned, By Deep Space Unmanned Probes, Which Further Serve, As Nuclear Fusion Power Plants, To Energize The Rings, And Monitor Black Hole Field Stability, At Their Induction Core: This Arrangement, Can Also Determine, Viable Distance, Of Potential Travel, Per Plasma Toroid, "Stargate Sets", A Series Of Stargates, Can Thus Be Positioned, To Leap Frog, Through Quadrants Of Space, As Required, Based On Their Generated Quantum Potential, Power Factor: Furthermore, These Toroid Stargates, Can Be Positioned, Along Non-Linear, "Path Roots", By Pitching Their Navigational Attitude, Like Lazar Reflecting Mirrors, To Reach Desired Destination Points, Not Unlike A Highway System - Satellite/Space Station, "Way Stations", Can Also Be Established, For Further, Scientific Monitoring, Of The Stargate System, These Stations, Should Be Built, In Curvalinear Sections, Forming, A Rotating Axis Ring, Upon Assembly, Creating, 1G Force, Of Earth Gravity, Within The Space Complex, Simulating Planetoid, Rotational, And Environmental Conditions - Provisions Made, For Biometric Circadian Rhythm's, As Well As Obvious Essentials, To Sustain/Prolong Life - The Influence, Of The Star Gate Array, Upon Relativistic Time, Will Effectively Slow Down, Or Perhaps Eliminate, The Human Aging Process, Which Is, Time Contingent: And Can Be Partially, Or Totally Arrested, Within Spacio-Temporal, Anomalous Zones
43) A Phantom Energy, Plasma Form, Like A Photographic Negative, Analogous Matter, Constellation In Hyperspace, Replicates, The Dematerialized Object, In Precisely Registered, Euclidean Vertices, Regardless, Of External Acceleration Imposed, Exerted, In Actuated Thrust, Upon The Quantum Event - This Holistic Conformity Of Structure, Independent, Of Apparent Material Form, Is Electromechanically, Field Conditioned, And Tied Inherently, To Biometric Structures, That Themselves, Carry A Low EMP Valence, In A "Weak Field" Envelopment - Therefore, The Quantum Energy, Transformed, "Anti-Matter Phantom", Of The Teleported Material/Biological Object, Is Projected Uniformly, Through Time Warped Linking, Telemetric Vectors, As A Photonic/Holographic Constellation: Reappearing, In Tact, Down To The Subatomic/Genetic Level, In Total Rematerialized, Decelerated, 3D Stabilized, Re-Solidified, Re-Materialized Form -
44) Starting And Stopping, A Quantum Leap Event, Attempted, In A Stabilized Toroidal Induction Field, With Energetically Matched, Black And White, Homeostatic, Worm Hole Sets - Requires Zero Point, Simultaneity, During Acceleration And Deceleration, Of The Spacio-Temporal, Displacement Event - Any Potential Power Flux, Will Alloy, Otherwise Independent, Quantum Phantom, Proximal Energy Fields, Into Haphazard,

Molecularly, Fusion Bonded, Conjoined Material, Causing Catastrophic Results, Upon Organic Matter - When Molecularly Bonding, With Inorganic, Proximal Matter
45) As In The Case, Of The Thermodynamically Infiltrated, Gyroscopic Nuclear Powered Flywheel, In Vacuo - Any Friction Point Of Contact, Introduced, Will Cause Microshifting, Of Rotational Telemetry, At The Axis Points, Out Of Symmetrical, "True Spin" - Causing Them, To Produce, Exothermic Breakdown, Thermodynamic Entropy, At Said Axiomatic, Pinch Points - The Gyroscopic Flywheel, @ Longitudinal Axis, Will Subsequently Experience, Uneven Rotational, Interval Magnetic Bucking Rates
46) With Central Centrifugal Elements, Moving Faster Than, The Magnetic Axis Termini This Screwing, "Spiral Action", Will Almost Instantaneously, Torc The Vertical Axis To Thermodynamic/Mechanical Stress, To Moment Of Sheer - Additionally The Rise In Reactor Core Temperature, In Vacuo, Will Heat The Fusion Chamber To The Point Of Generative Inefficiency, Producing Less Energy, Than It Takes In, To Operate/Function -
47) Therefore, The Entire Nuclear Flywheel Device Apparatus, Must Be Anti-Gravitationally Levitated, And Rotationally Stabilized, Without **ANY** Friction Points: Creating, The "Heliocentric Paradigm", Of A Perfect Balanced Rotation, Of A Synthetic Star - This Should Appear, As If Still, When Perfect Hypracceleration, Is Attained
48) The Same Parameters, Hold True, For Teleportation, Quantum Leap Events: Any Thermodynamic Breakdown, In The Mechanical System, Or Propulsion/Deceleration Rate, Will Cause Chaos Events, In The Improperly Energized, Dematerialized Constellation, Anti-Matter Field - In Both Instances, The Thermodynamically Afflicted, Portions, Of The Teleported Object, In Question, Will "Virtually" Tether To The Third Dimension, As The Rest, Of The Unaffected Object, Warps Into A Quantum Leap, "Phantom State" - The Tethering Point, Of Entrophy, Becomming An Achilles Heel, To The "Disturbed Reaction": Which Will Likewise, Take The Path Of Least Resistance, In Conservation Of Energy, And Conjoin, Will The Closest Material Object/Solid Matter, In It's Localized Proximity - Creating, A Domino Effect, Of Nuclear Fusion Events, In The "Rematerialization Process" - Teleported Incoming Rematerialized Matter, From Quantum Events, Will "Stack" Like A Multiply Vehicle, Car Accident, Occurring, At The Speed Of Light - Becoming Entangled, And Atomically Trapped, At The Collision Nexus, Of The Thermodynamic Entropy, @ "Bleed Point"
49) Another Danger Factor, Results When Magnetic Polarization, Is In Reaction Compromised: No Longer Acting, As A Repulser Field: Instead The Object Becomes An Induction Vortex, Opportunistically Targeted, And Crushed Down, To The Subatomic Level, Taking On, The Gravitational Load, Of The Surrounded Effected Environment - Super Compressed, In Conformity, With A Euclidean Solid, Vertices Converge, In Cubed Facia State, Of Gravitationally Bombarded, Hypercompression: Crushed, Like A Diamond, From A Ton Of Coal, Yet Still Carrying It's Original, Gross Weight: Like An Imploded Black Hole Sun, Reduced To The Size, Of A Basketball -
50) Aforesaid, Stargate Portals, Can Also Serve, As Super Computing/Transmission, Quantum Conduits, Offering An Ideal, Time Suspension/Displacement State: To Create, An Ultraconductive Medium, For High Frequency Transmission - Linear Accelerators, Propelled By Frictionless, Side Mounted Atomic Flywheels, Act As Signal "Transmission Amplifiers", Taking Any Introduced Particle, And Turning It, Into An Anti-Particle, Traveling Beyond, The Light Barrier - This Is Particularly Effective,

When Three Separate, Particle Beams, Moving At Different Rates, Of Frequency Oscillation, Are Aimed At, A Triangulated "Strike Point", At The Black Hole Epicenter A Direct Current Power Frequency, With Near Flat, Line Oscillation Characteristics, As A Particle Beam Constant, An Alternating Current Wave Form, In The Theta Spectrum, Serves, As A Carrier Wave, Exciter Frequency, And Helmholtz Induction Coil, Emitting A Microwave/"Chopped Frequency", All Connected To Each Other, And The Central Plasma Fusion Reactor, Open Delta, In A "Dead Short" Electronic Configuration This Particle Entanglement, Helps To Hyperaccelerate, And Time Displace Objects Sound, Can Also Be Stepped Up, And Transmitted, Using Resonant Cavity, "Sonic Canons", Which Can Be Used, To Calibrate, And Distance, Stargate Toroid Sets - Calculating Modular, Sonic Distribution, Along The Teleportation, Portal Distance - All Of These Techniques, Can Be Empirically, Field Tested, For Optimal Output, And Tuning Efficiency, With Thyristor Arrays, Set To Create, Maximum Excitation, In The Quantum Field, As Well As Serving, As A Curcuit, Impedance Ballast, To The Controlled Electronic Short - Ideally Achieving, A 26% Induction Rate, At Maximum Generation, Valence Output: As Well As Stabilizing, The Quantum Field, During Teleportation, Spacio-Temporal Displacement, During Super Computing/Communication

51) Additionally, Faraday Cages, Can Be Target Positioned, To Induct/Trap Escaping Anti-Matter, Before It Reaccelerates, Back Into The Relative Space - As It Exits, The Tokamak, Nuclear Fusion Powered, Slush Hydrogen, Superconductive, Toroid Ring Stargates - Stored Antimatter, In A Homeostatic, Electromagnetically Stabilized Form, Can Further Serve, As A Fusion, Nuclear Fuel, Breeder Reactor, For Anti-Particles: Centrifuging, In The Hydrogen Slush Superconductor, Toroid Ring Exciting, The Cyclonic Flywheel Hyper-Rotation, Of Nuclear Particles, Exponentially, To Quantum Warping Threshold - Producing A Stable, Tachyon Field, Black Hole Located, In The Precise Center, Of The Stargate Ring, Energized Slush Toroid, In/@, The Epicenter, Of The Temporal Displacement, Space Field, Warping Event -

52) "NicoHaben" (= German), "Hobbled" Three Way, Synergized Frequencies, Crippled Man, With A Cane = One "Good Leg" + One "Hobbled Leg" + One Thin, "Cane"/Support, = "3/4D, X Frequency" - Target Entangled, By Open Delta Array, Of Three Linear Accelerators, Striking A Point, Of Singularity, At The Toroid Stargate Portal Alternating Current, (AC) Direct Current, (DC) Independent Power Sources: With A Helmholtz Induction Coil, Frequency Modulated, By An Oscillation Tuning, Thyristor Array - Will Geometrically Increase, Warp Potential, And Therefore, Anti-Matter Gross Yield, In The Magnetic Faraday Storage, Containment Devise

53) Correctly Harvested, And Containment Sealed, In A Hermetic, Magnetic Bottle Array Stabilized Anti-Matter, Can Be Utilized, As A Nearly Inexhaustible, Power Source/ Supply - With Tremendous, Valence Dynamics, Hitherto Unheard Of - Designed With Massive, Graduated, Step Down Transformer, Dynamo/Farm, Equipped With "Choke Coils", Along Super Cooled, Bus Bar/Curcuit/Power Lines, Prevent Power Inversion Factor, "Backwash", Which Will Implode, The Generators, If Not Choke Coil, Failsafe Installed - To Prevent, Power Reflux Implosion, At The Central Power Source - These Anti-Matter, Fueled Dynamos, Can Provide/Supply, The Power/Electrical Needs, Of Entire Planet, With Deep Space Harvested/Teleported, Trapped, Zero Point Free Energy - After Power Conversion, Through, Ohm Resistance/Work Load, The Anti-Matter Energy, Reaccelerates, Back, Into Relativistic Hyperspace -

54) Additionally, Making Off World Colonization, A Population Explosion, Imperative - Theoretically Viable, To Meet The Extreme Energy Requirements, Of Off World Construction: Complex Tunneling, Water Purification, Atmospheric Creation/Terra Forming, Magnetic Shield Regulating, Exospheric Atmospheres, From Solar Radiation Bombardment, Deflection Of Extreme Heat, In Thin Atmosphere: A "Magnetic Umbrella" - This Would Also Control, Liquified Water Source, Evaporation: During The Process, Of Polar Ice Cap Melting, With Fusion Reactor, Plasma Canons

55) Primary Location Points, For Stargate Sets, Should Be Near, The Closest Planets, In The Solar System, That Exert The Least, Gravitational Pull, Or Electromagnetic Instability, And Have The Best Prospects, For Terra Forming, And Colonization: Mars, Has Polar Large Polar Ice Caps, What Appears To Be, The Remains Of A Canal, Or River System, Indications, Of A Thin Atmosphere - A Fusion Powered Automated Nuclear Probe, Designed To Melt Polar Ice, Capable Of Heat Generation 12X+, Magnitudes, The Temperature, Of The Suns Core, Would Begin, The Process, Of Releasing, Trapped Water Vapor, And Oxygen, Into The Martian Atmosphere: Plant Exobiology, Extracted From, Ice Comets, And Other Hardy Forms, Of Vegetation Can Be Seeded, On The Surface, Once The Polar Caps, Are Partially Liquified - The Reintroduction, Of An Atmosphere, As Well As, A Liquified Water Source, Would Regulate, Surface Temperature, Of The Martian Day/Night Extremes: The Terra Formed, Plant Species, Would Be, A Perfect Barometer, For Biological Viability, On The Surface, As Well As Being, An Additional Oxygen Generating, Atmosphere Yield Production Methodology - Targeted Around, The River/Canal, New Water Tables - This Would Be, An Ongoing Process, Until Mars Reaches, Earth Atmospheric Parity - In The Mean Time, Subterranean, Airlocked Structures, Can Be Constructed Fusion Reactor Powered, Automated/Robotic, Mole Boring Machines, Can Be Utilized, For Plasma Wave, Smooth Boring, Of Extensive Bedrock, Tunnel Systems - Terra Formed Plant Biology, Can Be Checked, For Genetic Mutation, Due To Short Duration Crop Cycles, These Biometric Anomalies, Can Thus Be, Quickly Detected - Before Moving On, To More Complex, Life Form Introduction: Caged Insects, Avians And, Laboratory Rodents, Seem The Best, Initial Test Subjects, For Genetic Analysis

56) The Genesis Reintegration Point, Of Rematerialized, Hyperaccelerated Matter, Is Governed By, A "God Particle", Genetic Structural Template, Which Reorganizes Itself, Like A DNA/RNA, Zygomatic Encodement, Back Into It's Original Organic, Or Material Form, This Structural Memory, Is Innate, In All Three Dimensionally, Conditioned Matter, After Leaving Hyperspace, Teleportation Events: Through A Energetically Stabilized, Tokamak Powered, Plasma Ring Toroid Set, At Perfectly Calibrated Proximity - This Occurs Naturally, In Special Event Quadrants Of Black Hole, While Hole Space Nexus Points, It Is Merely The Scientific Issue, Of Synthesizing, These Conditions, By Which Such Physics Occur - The "Pi Factor", Interdimentional, Event Horizon, Will Naturally, Seek Out Equilibrium, In Power Factor Pairing, The Energy Exchange, Between Third And Fourth Dimension, In This Manner, Is Ubiquitous, And Serves, As A "Eternal Warp Drive" In A Perpetual Engine Network, Which Energizes, Expands, And Creates New Matter, In The Ever Expanding Universe, At The Increasing, Speed Of Light: From The Big Bang, Where Stabilized Matter, Was Originally Formed, In Plasma State, To Progressive Ossification, And Compressing Of Matter -

To Eventual Return, To Warp Drive Dematerialization, Of Said Matter, In Thirty Billion Year, "Universal Life Cycles": Where The Material Universe, Of The Third Dimension, Implodes Back, Into Dematerialized Pure Energy: Where, At A Certain Valence Point, Of Dynamic Surplus Compression, It Shunts Energy Back, Into Physical Space, In A Subsequent Series, Of Big Bangs, Throughout Eternity - First, Photon Radiation Particles, Created, Then Gaseous Plasmas, Then Crystalized Electron Chains, Forming Cosmic, Electromagnetic Dust - These Consequently Constellate, Into Elemental Isotopic, Molecular Compounds, Eventually Forming, Chemical States, That Are Conducive, To Generative Biological Life Forming - Big Bangs, Also Occur, In Decentralized Space, At The Formative Edge, Of The Universe, Creating, A Surrounding, Energetic Encapsulating Sphere, Where The Internal Black Hole, White Hole, Worm Hole Sets, Mine Energy, From This General Fusion Field, Which May Exist, On Multiple Planes, Of Magnitude: On A Macrocosmic Scale, Too Large, To Physically Calculate - All Seeming, To Be Designed To Achieve, The Greatest Coefficient, Of Energy Yield, Through, **Mass, Times** (**Acceleration, Squared**, At Ten, To The Sixty Fourth Power - Like The Archimedean Lever, Upon, A Tipping Scale, Pendulum Swaying, To The Terminal Acceleration Potential Point, En Masse - Gravitating, Toward The Energetic, Outer Periphery, Of Newly Created Space, In Omnidirectional Expansion, Lagrangian Field Dynamics, Form, A Central Genesis Event - Light Acceleration, Through Quanta Of Space, Create Paradigm Shifts, Which Recondition Matter, Create Linear Time: Evolutionary Radiation, In Biology, Cause Environmental Extinction Cycles, To Occur, Consistent, With The Darwinian Theory, Of Survival, Of The Fittest - This Seems, To Hold True Also, On A Cosmological Level, In An Upwardly Ascending, Evolutionary Spiral - Light Radiation, Duration Cycles, Time Elapsing, In Space, Seem The Causative Factors, In Paradigm Shift Phenomena: And Seem To Carry, Morphogenic Resonance, As Postulated, By Sheldrake - This Evolutionary Model, Of Biomorphorphic Mutation, Can Potentially, Be Excited, Or Accelerated, By Hyperspace Teleportation, Creating, A Quantum Leap, In Human Evolutionary Development, By Such A Theoretical Line, Of Postulated, Deductive Reasoning - Since The "Modulating State" Light Source, In Question, Has Been Accelerated Beyond, The Conventional Time Line, Parameters, Into A Future Trajectory, Potential Evolutionary, "Upward Curve" - Human Chrononauts, May Well Develop, Precocious Evolutionary Genetic Benefits, In Such, "Paradigm Shifting", From A Current Event Horizon, Quantum Leap Perspective, This Seems Highly Probable - From A Historical, Epistemological, Biologically Advanced, Evolutionary Vantage Point

57) The Phenomena, May Also Occur Inversely, With Time Travel, Into The Past: Therefore, A Psychological, Zero Point Time Referent, "Attunement", May Be Required, To Readjust "Chrononauts" Who Have Ventured, Into The Distant Evolutionary Past - Fortunately, Human Evolutionary Development, Requires Thousands Of Years, To Express Itself, In The General Population - Short Term Exposure, To Past Time, Should Have, A Minimal, "Echo Like Effect", And Eventually Dissipate, Upon Return

58) The Anti-Grav Levitating, "German Bell" Has The Advantage, Of Having The Aforementioned Technology, Self Contained And Modular - Incorporated, Into It's Design; The Toroid Plasma Ring Array, Situated At The Base Of The Bell, Is Pendulum Weighted, And Symmetrically Trued, For Balanced Vertical Ascent - The Ring Base, Is Reservoired, Through The Curvilinear, And Graduated/Weight Balanced Hull, Creating

A Gyroscopic, Hydrondynamic, Venturi - Attached To, A Fusion Nuclear/Electronic Terminus, At The Balanced Crown, Of The Devise: Energizing A Central Induction Axis, Which Bisects, Diaphramatically, Into An, "Magnetic Isolation Field Transformer", To Excite Magnetized, Zero Point Energy Induction, In The Lower Half, Of The German Bell The Central Modular Axis, Furthermore Supports, A Twin, Rotationally Variable, "Jacobs Ladder" Tesla Coils - These, In Lieu, Of "Helmholtz Coil" Induction, Tower Span Wire, Configuration - The Base, Of The German Bell, Utilizes A, "Sky Iris", An Adjustable Aperture Induction Port, To Thyristor Modulate, Free Zero Point Energy Induction, From The External Environment, Into The Internal, Central Ring Core, During Tachyon, Black Hole, Anti-Matter, Formation: Regulating The Reaction, And Controlling, Anti-Gravity Escape Acceleration, Upon The Gross Mass, Of The German Bell Apparatus -

59) The Superconductive Plasma, Generates Trumpeted Funnel, Creating A, "Whirlpool Venturi", Hydrodynamic Plasma, "Archimedean Screw"/Chemical Flywheel Rotation, Agitating A Flushing/Sumping Effect, Into The German Bell, Lower Flywheel, "Slush Hydrogen, Ring Chamber", By A Series, Of Graduated Venturi, High Compression Valves, Which Establish Ionic Direction Flow, In A Vertical Line, Of Ascent And Descent - This System Helps Induce, Flow Agitation/Conduction/Induction, Of Super Cooled, Slush Hydrogen Plasma - At Controllable And Variable, "Flood Gate" Rates, To Further Refine, Navigational, "Lift Dynamics" - Nuclear Power External Gyroscopes, Posted At Euclidian, Directional Vertices, Can Selectively Alter, The Trajectory, Of The German Bell, To Any Flight Attitude, When Sufficient Alleviation Of Gross Weight, Is Neutralized, In Gravity: To Create Zero Point, Levitation Profiling

60) German Bells, Are Ideal Propulsion Systems, For In Atmosphere, Traction Systems And Large Scale, Towing Operations: No Longer Contingent, Upon Derrick, Or Jib Mechanical Engineering, Tower Arrays, Or "Terminal Weight", Constriction Load Parameters - Furthermore, Effectively Replacing, Goddard/Parsons/VanBraun, Thiokol, Solid Rocket Fueled, Multiple Rocket Staging, With Further Capacity, To Maintain Geosynchronous, Levitation Positioning, While Sustaining, Tremendous Workload Torc: By Neutralizing, Surrounding Gravity, Upon Itself, And It's Cargo - Further Efficiency Enhanced, By Weight Reduction, Of Not Having, Combustable Fuel To Carry, Substantially Lessening, The Co-Efficient Of Drag, Upon The Vehicle

61) Like The Norse Mythology, Referencing, Thor: The Reincarnated Soul, Of The "Dying God" Balder, Killed By The Trickster God, Loki, With Envenomed, Mistletoe - Odin, Searching, Upon The Earth, In Sacred Journey - Finally Found, By The All Father, Wotan, Now In Human, Mortal Form: Disguised, As A One Eye, Sturdy Rogue - Festooned, With A Wide Brimmed Hat, Worn Askance, To Cover His Deformity - Father Aesir God, Comes Upon, A "Lamed Physician", With A Walking Staff: In His World Travels, Which Take Him, To The Far East, To The Original, **Migration Point**, Of The Indo-Aryan, Germanic Peoples - Recognizing His Spirit, To Be That, Of The Resurrected God, Balder: Odin, Instructs, This **"Hobbled"** Broken Man, This "Lamed Physician", From The Far East - To Strike, His Walking Staff, Upon **"Earth Ground"**, By Such, "Geomantic Process", He Is God Transformed - The Walking Staff, Turned Into, War Hammer, Becomes An, **Accelerating Flywheel**, Providing, **"Levitating Propulsion"**, And Tremendous Locomotive, **"Destructive Kinetic Force"** When Flung, Always Returning, To Thor, With **"Iron Magnetized"** Gauntlet - When Spun, "Centrifugally", Above The Head, Of The "Awakened God" Of Thunder, "Inducting

Lightning", Unto Himself: By The "Swirling Hammers", Hyperconduction: = "Lightning Rod" - His Hobbled And Broken, Lamed Mortal Form, Is Metamorphasized, Into The Aesir: Norse Barbarian, Tribal Pantheon: God Of War, The "Resurrected Balder"
62) The One Eye, Of Odin, In Mortal Disguise, Is The **"Induction Portal"** = **"Sky Iris"** The **Lamed And Hobbled**, Physician From The Far East = Is The **"Nikohaben"** = "Lamed/Hobbled" Triform, "Multi-Oscillation", Synergized **"Carrier Frequency"**: That Breaks, The Light Barrier, When Amplified, Through, Linear Particle,, Three Phased, Hyperacceleration - The **War Hammer Spun, Centrifugally,** Becomes, An **"Antigravitational Flywheel"**, Alleviating It's Own Gravity, Repulsing That, Of The Surrounding Environment - Inducting, Zero Point Energy, Of Scalar Ball Lightning - From Atmospheric, Zero Point Energy, As Dark Matter - A Black Hole, Stargate Nexus, In The Central Epicenter, Of The Rotational Axis, Of The Flywheel: Where The Interdimensional, **"Pi Axis Point"**, Between Third And Fourth Dimension: Forms, Like The Asgardian Rainbow Bridge, Of Bifrost, Spanning The Universe, Through A Series, Of Interconnected, **"Worm Hole, Valence Match Sets"**
63) The Theory Of Rotating Stars, On A Paradigmatic, Axiomatic Level, Is The Guide Line, Of Logic, And Underlying Principle, Of Unified Fusion Field Theory, Of Simultaneity - The Fusion Physics Expressed, Are That, Of A Perfectly Rotating Star, With, A Stable Rotational Axis: No Sidereal Sway, Gravitational Wobble, Sidereal Tangent, Of Rotational Offset: A "Completely Stable" Spheroid, Weighted And Balanced, In Closest Conformity, To Geometric Pi: Producing The Most Stable And Productive Gravity, And Valence Output, With The Least Amount, Of Wasted Thermodynamic Breakdown - The Condition Of The Star, In Various Phases, Of It's Cosmological Life Span, From Gaseous, Amorphic Plasma, To Mature Star, To Expanding Giant, To Imploding Dwarf, To Black Hole Sun - All Indicative, Of Phases Of, **Mass, Times Acceleration Squared** - And Are Physically, Replicated, Insitu, In A Tokamak, Fusion Reactor, At Various, Acceleration Shift Rates - Initially Creating, Earth, Gravitation Field Weakening, Upon The Reactor Physical Plant, To The Productive Point, Of Gravitational Pull, Neutralization/=A/G Levitation, Production, Of Upward Thrust, In Proportion, To Increased Acceleration Rates, Of The Flywheel: Gyroscopic Dimensions, Metallurgy, Gross Weight, Are So Factored Thyristor Controlled, Rate Of Flywheel/Toroid Spin, Can Cause Multiple Navigation Characteristics: Including Ascent, Hovering, Descent, Similar To Helicopters, But With Much Higher, Acceleration Rates - The Magnetic Shielding, Of The Vehicle Also Relieves, Aerodynamic Lift, And Drag, As Well As Positive, And Negative Wind Resistance, No Longer Factored, Into The Design Equation - Wings, In This Particular Instance, Are Impractical - Solid, Liquid Or Gaseous Fuel, Is Eliminated, Along With Extra Load: As The Power Plant, Is Nuclear Fusion Powered, In A Closed, Plasma Ring, Or Vacuum Chambered, Gyroscopic System
63) Precise Attention To Detail, Must Be Applied, To Manufacture, And Super Alloying, Gyroscopic Flywheel, Apparatus - No Matter How Large: The Devise, Is Designed To Be, It Must Have, A Perfect Molecular Consistency: Spun Foundry Casting, So That It Will Withstand, Huge Centrifugal Physics, And Perfect Balanced, And Symmetrically Weighted, Throughout, The Devise - Like The Truing, Of A Jet Turbine, Or Airplane Plane, Or Ship Propeller: Any Deviation, From "Trued Symmetry", Will Cause The Flywheel, To Generate A Wobble, Which Will Geometrical Increase,

To, "The Moment Of Shear", Weight Distribution, Miscasting, Will Cause Super Alloyed, Aerospace Grade Materials, To Accelerate, At Different Rates, Throughout The Physical, General Dynamics, Of The Devise: The Uneven Mass, Will Cause, Cast Portions, To Be Become, "Virtual"/Holographic/Interdimensional, At Hyper Acceleration: Causing The Underaccelerated, Remaining, Solid Matter, To Fly Apart, At The "Virtual Seams" - Any Manufacturing Defects, Along, These Lines Of Stress, Must Be Quality Controlled, To Avoid Turning, Such High Speed Apparatus, Into A "Ballista/Grenade" - Failsafes, To Assure, The Most Stable, Rate Of Rotation, Perfect Weight Distribution, Frictionless Bearing Points, And Mag/Lev, "Truing", Of Vacuum Rotation, Are Therefore Instrumental, To Device Efficiency - Temperature Fluctuations, In Vacuo, Are Also, To Be, **Absolutely Avoided**, And Will Cause, Thermodynamic, Entropic Breakdown, In The Reaction: Making It, Unproductive - The Mass Equivalent, Of A, "Dead Short", In The Valence Output/Acceleration, Of The Flywheel: Any Contact Point, In The Rotational Field, Will Serve As Brakes, And Stop The Reaction Production - Any Physical Contact, At Hyperacceleration Rates, Will Cause Super-Heating, Immediate Thermal, Wear And Tear, And Eventual, Implosion/ Explosion, Once The Super Cooled, Vacuum Chamber, Has Been, Absolute Zero, Space Temperature, Heat Compromised - Cold Nuclear Fusion, Can Only Achieve, Productive Yield, Under These Interlinking, Closed System Parameters, To Produce And Induct, Zero Point Energy, @ Generative Power Factor

64) The Amalgamated, Accumulated, Locomotive Force, Increasing It's Own Gravity, In Increasing Rate, Of Gyroscopic Spin: Like A Pail, Full Of Water, Attached To Rope, Spun In Rotation, By A Child: None Of It Leaving, The Gravitational, "Containment Field", As Long As The Child, Provides Sufficient, "Spun Motive Energy" - To Keep The Bucket, Sufficiently Accelerated - So The Pail Contents, Will Not Spill Outward - Theoretically, If, The Child Were Strong Enough, The Water Load, Of Rope Tied Pail, "Sufficiently Heavy", To Create, A Powerful Accelerating Flywheel, The Child, (Power Plant) The Pail And Rope (Toroid Flywheel) Would Relieve Their Own Earth Bound, 1G, Of Load Gravity, And Begin To Levitate - (Anti-Gravitational, Solar/Planetary, Autonomous Magnetic Counterforce) - The Induction, Of Free Zero Point Energy, Excited By Di-Polarized, "Warring" Magnetic Fields, In Dynamic Repulsion, Creates "Invisible" Locomotive Force, Being Part, Of The Surrounding Natural Environment: It Cannot Be Detected, In The General Field Quantum, Except, For The Increased, "Momentum Acceleration", Co-Efficiency Profile, Of The Hyperspun Object, As It Geometrically Relieves, It's Own Gravitational Weight, Countering Drag And Torc, In The Process - While Increasing It's Own, Gravitational Characteristics, To The Point Of, "Infinite Mass Acceleration", Creating, Spacio-Temporal Displacement, In It's Unique Gravitational, Autonomous Energy Field, In A Lorentz Invariant, Inverse Levitational, Vs. Newtonian Trigonometry, Fall Rate Descent Table - In Accordance, With An Anti-Gravitational, As Opposed To, Gravitational Physical Events: The Physics, Are Consistently, Inverted, Proportionately Opposite

65) Two Opposing Magnets, Juxtaposed, Gyroscopes, Share This, Weight Alleviating Levitational Characteristic, Foucault's Pendulum, Another Devise, For Ascertaining Earth Gravitational Latitude, Is Also Effected, By The Lagrangian, "Magnetic Flux", Surrounding, Rotating Magnetic Spheroids: As Well, As In A, Common Compass, Detecting Magnetic, Polar North - When, The Flywheel/Gyroscope, **(Equator)** Is Magnetically "Bucked" By Opposing Magnets, On The Flywheel, **(Rotor)** And On

The Central, Structural Armature **(Gimbal)** Pulse Energized, And Microwave Regulated, Like A Machine Gun, A "Tuning Thyristor", Regulating, "Pulse Intervals" In The Case, Of Slush Hydrogen Toroids, Nuclear, Ionic Flow Stream, Controlled: Like, Inserting, Or Extracting, Plutonium, Fuel Rods, In A Fission Nuclear Reactor: Rates Of Acceleration, With Navigational Analogs, Are Then Possible, In The Anti-Gravitational, Levitation, Temporal-Displacement State: Productive, In This Manner

66) Side Mounted, Mag/Levelated, Linear Accelerators: With Cored Out/Rifled Axis, Essential Replace, The Fermi Cyclotronic, Or The Einsteinian, "Cauducean Coil" Variant: With A, "Barrel Length", Of Four Feet, Or More, When Hyperaccelearted, By Microwave Bucking Coils, Along The Flywheel Rotor/Gimbal Array, Creates Particle Introduced Acceleration, At Simultaneity: Physically Existing At Two Ends, Of The Barrel Array, At Once: This Space/Time, Anomalous Phenomena, Which Thus Refutes, General Relativity, On It's Face: It Is Essentially, A Rudimentary, "Basic Teleporter" Array, Which Magnetically Amplifies, And "Thickens" Introduced Matter, Into A "Focusing Coil", "Particle Canon" - Weak Lazar Emissions, Can Be "Broadcast" Through, The "Quantum Tunnel": Geometrically Increased, @ Terminal Warp Speed @ Exponential, "Power Factor Level", Of Valence Output - From Microsurgical Operations, To Sky/ICBM Defense: It Also Creates, The "Magnetic Bubble", And Nuclear Triggering Particle, To Initiate, A Productive, Nuclear Fusion Reaction: The Cyclonic, Magnetic Profile, Not Unlike, An Archimedean, **"Hydraulic Screw"**, Magnetically "Bundles", And Venturi, Energy, Into A, Synergistic, "Tethered Spiral" - Isotope Striking, Fissile Material, At The Speed Of Light: Within, The Magnetically Shielded, Doubled Hulled, Tokamak Chamber: Kept Cool, In Vacuo, By Gamma Radiation Fallout, Between The Reactor Hulls: Contained, By External, Tremendous Magnetic, Repulsive/"Containment Force" - The Side Mounted Linear, Horizontal Accelerator, Must Be Levitated, And Rotationally Stabilized, By Linear Arrays, Of Frictionless, Magnetic Yokes, To Achieve Productive Simultaneity, "Strike Efficiency"

67) Bucking Coils, Microwave Energized, In Geometrically Stepped, Acceleration Pulse/Repulse, "Shunting Intervals", Magnetic Pitch And Attitude, Aligned On Rotor And Gimbal, Should Be Aligned, Unidirectionally, With Greatest Angled Interface, To Aid In, The Pitch And Direction, Of Cyclonic Rotation - Bucking Coils, Should Only Exhibit, Opposing Magnetic Characteristics, When They Reach Ideal, Interface Passing "Bucking Trajectory" - This Must Be Calculated, By A Supercomputer, To Regulate Microwave, "Timed Momentum Pulses", Delivered, In Nano Seconds, And Beyond - The Device Rotation, Itself, Can Serve, As "Switching Gear" To "Inform" The Electronic Distribution System, Of The, "Core/Gimbal Telemetry, Nexus Point", At Optimal Rotation, "Bucking Repulsion Point" - Much Like A Distributor Cap, With Contact, Spark Plug Points, In Sequential Firing Array, And A Spun, Contactor/Rotor, To Complete, The "Rotating Switching Circuit": When The Motor Pistons, Are In Upward Piston Chamber, Fuel Firing, Optimal "Engine Head Positions" - To Coordinate, The Opposing Set Of Pistons, The Case, In A "V" Type Engine, Crank Shaft, Actuating The Common, Gasoline Powered, Combustion Engine - But, On Super Efficient Level

68) The "Open Delta" Electronic Array, Throughout, The Tokamak/Linear Accelerator/ Double Hulled, Fusion Reactor, Magnetic Containment Field, Must Be Ballasted, In Homeostatic Equilibrium, As The Fusion Reaction, Increases And Decreases, Unilaterally, By Thyristor Controls - Otherwise, The Reaction Collapses, And

Becomes, Energetically Unproductive: In Part, Serving, As A, "Failsafe Mechanism", For An Unstable Reaction, Nuclear Fusion, The Magnetic Field Bubble Collapse - Which Is Gravitationally, Rotationally, And Magnetically, Interdependent - Expressed Physically, By Mass, Times Acceleration, Squared - Additionally The Induction, Quantum Influx, Of Free Energy, Is Critical, In Making The Power Factor Level, Within The Reaction, Substantially Higher, Than The Energetic Input, Reaction Consumed: Like The Theory, Of Rotating Stars, In The Vacuum Of Deep Space, Or An Incandescent Lightbulb Filament, Glowing, In An Inert Gas Filled, Bulb/ Chamber, The AC/DC Oscillation Cycle, Conditions, Of Solar Astrophysics, Must Be Technologically Replicated, And Stabilized, Up And Down: The Power Factor Scale, Proportionate, To The Amalgamated Mass, Involved, And Synchronized Rotation Rate, Of The Sustained, Nuclear Fusion Reaction, Happening Within Tokamak Flywheel Array: To Relieve, It's Own Earth Bound Gravity, General, Mag/Lev, Anti-Grav, "Solar Properties", And Physical Weight/Time/Space Dynamics, Into Eventual Simultaneity

69) Hysteretic Damping, Vibrational Oscillation, Friction Energy Loss, If Often The Major, X Factor, In Energy Reduction Scenarios - The Least Efficient Traction Methodology, Acceleration, Of A Wide Inflated Tire, Upon A Variable Density, Pliant Vibrational Surface - The Second Magnitude, Of Traction Economy, Comes From The Initial, Conversion "Motility Point", From Static Inertial Resistance, To Rolling Point, Induction Of Productive Movement, Overcoming Initial Resistance Factor, @ It Greatest Power Factor, To Rolling Motility, "Exciter Point": Static Inertial Resistance, To Rolling Motility, Overcome By Thermodynamically Conditioned "Rolling Resistance" = Which In The Greater Case, Of Locomotive Traction, Reduces, To Virtually, "Zero Point Non-Resistance" Once The Accelerated, Gross Amalgamated Mass, Autonomously Conditioned, In A General Dynamic, Forward Trajectory, Envelope - Creates Localized, Power Factor, Localized Field, Induction Of Free Energy, Within The Acceleration Sphere, Of Surrounding Physical, Magnetic/Gravitational/Environmental Influence

70) Spacio-Temporal, Displacement Anomalies, Of Low Valence Amplitude, Are Thus Discreetly Detectable, In Parallax Timed, By Measurable, Synchronization Incongruities: At Fixed Field Point, And Upon, The Accelerated Mass, Of The "Locomotion Enveloped" Motile State, Vehicle - Increasing Time Discrepancy, Widening, In Quantum Flux, Along Increased Distance/Acceleration, Of Travel Path

71) Further Increased, Within Lorentzean Constant, General Field Dynamics, As The Rolling Resistance, Drops To "Zero Point" Of Entropic Friction, Motive Resistance,

72) The "Interior Ball", Curved Bevel, Of The "T" Rail Head, Supplies, A Rigid, "Static Solid", Width Pre-Gauged, "Velocity Flux Conduit", Of Locomotive Traction, Further Reduces, "Hysteretic Damping", @ The Wheel/Rail, Interface, "Traction Point" - Providing, A Narrow, Static, Non-Oscillating, Thermodynamically Efficient, Cross Sectional Modulus, For Paradigmatic, "Rolling Point" Of "Locomotive Motility" - Subsequently, "Engineering Out", The Hysteretic Energy, "Traction Resistance", Thermodynamic Depletion, Inherent, In Wide Traction, Soft/Malleable/Uneven, Road Surfaces, Causes Diminished, Inefficient Acceleration, Traction Profiles -

73) Magnetically Levitated, "Bullet Trains" Further Eliminate, Static Inertial Resistance, By Levitating, The Gross Weight, Of The Train, Insitu - Like A Crane Lifting, A Heavy Object, By Cable Hoist, Easily Manipulated, By Ground

Workers, Into Intended Position: Suspended Weightless, In The Air - Thus Eliminating, Static Inertial Resistance, To Zero Point Efficiency - Rolling Resistance, Is Isolated And Eliminated, From Direct, Track Head Contact, Maximizing, Accelerated Propulsion Potential, To The Realm, Of Aerodynamic Lift - The Train Mass, Is By Propulsion, Amplified, By Anti-Gravitational, "Force Velocity", @ Three Hundred MPH +, Effecting, Initialized Black Hole Induction, Temporal Displacement Effects, Occurring @ Hyperaccelerated Mass, Inertia Path/Locomotive, Solid Static, "Right Of Way"

74) Static Resistance, Is Always Greater, Than General Dynamic, Rolling Resistance, Initiated By, Motile, Productive Traction Rates, Is Diminished, As A Negative, Co-Valent, In Geometrical Progression: In Relation, To Mass, Times Acceleration, Squared -

75) The Horizontality, Of The Locomotive Path, Of Trajectory - Gradually Contouring, Over, A Lagrangian Spheroid, Magnetic Planetary, Field Axis Rotation: Further Enhances, The Acceleration/Locomotion, Envelope: Not Required, To Fight "Gravitational Pull Vertically", Therefore, Using It, To It's Physical Advantage - Exceptions Occur, With Sidereal Pull, On Curvilinear Traction, And On Inclined Traction, Under Reciprocating Load, Gravitational Pull, At Degree Of Inclination - Downward Pitched, Trackage, Further Enhances, Thermodynamic Neutralization, "Breaking", Then Becomes, The Physical Issue = "The Runaway Train, Effect"

76) Elimination, Of Train Wheel, Truck Sets, And Their Axle, Friction Bearings, With Mag/Lev, Track Repulser "Bucking Trucks", Still Further Reduces, And Refines, Friction Energy Loss, In Zero Point Induction, Frictionless Locomotion = "**Fusion Flywheel**"

77) The Universe Itself, Operates, In Much The Same Manner - From The Big Bang, Of, Static Inertia Resistance, Energetically Overcome, To Productive Motility, Of Initiated, Rolling Resistance, At Primordial Base, "Genesis Level" Of Speed Of Light: As Matter Cooled, And Hardened, Gained Increased, Molecular Mass, And Atomic Weight, Forming Suns, Planets, Moons, Coalescing, Into An Astrophysical General Dynamic, In Ubiquitous Cosmogony - The Amalgamated Matter, Big Bang Hyperaccelerated, At Higher Rates, Of Gravitationally Relieved, Coefficiency: Increasing, The Speed Of Light, In The Expanding, Path Of Travel: From The Big Bang Epicenter, Inducting Black Hole, Dark Anti-Matter Energy, Along The Way - This Attends, All Matter, Hyperacclerated States, Of Quantum Flux: Induction, Giving Surplus, Power Factor, To The Ever Expanding, Universe: Like a Perpetual Engine, Or A Tokamak, Fusion Reactor, On A Macrocosmic Scale - And Therefore, **REDUCING**, Not **INCREASING**, Thermodynamic Breakdown, As Has Been, Erroneously Speculated, In "Relativity", And Other Twentieth Century, Physics Theories: Before, The Development, And Experimentation, With Nuclear Fusion

78) Thus, A Basic Schematic, Is Drawn, Of The Fundamental, Quantum Fusion Matrix

79) General Field Dynamics, And Their Solar System, Galactic, Macrocosmic, White Hole, Black Hole, Matched Interactive Counterparts, As Being Part, Of A Ubiquitous, Magnetically Tethered, Plasmatic Flux Field, Expanding @ Infinite Potential And Increasing, Geometric Magnitudes, @ The Progressively Variable, Speed Of Light - In Approximately, Thirty Billion Year, Acceleration Cycles, That Continue To Expand, From Big Bangs: Matter Production, To Warp Speed, Paradigm Shift, Into Pure Energy, Anti-Matter Quantum Fields: This Occurs Continuously, In Various Quadrants, Of Specialized Event, Quantum Conditioned Space, And Is, Hyper Dynamically Reciprocal, In Simultaneity, Unconditioned, By Physics Of Localized, 3D Time Space -

Understanding, And Utilization, Of These, Worm Hole Sets, Are The "Invisible", Fusion Physics Mechanics, That Drives The Universe, The Numberless Stars, In Galactic Formation, In The Space Cosmology, Exert Electromagnetic Influence, Upon Each Other, Like The Nuclear Particle, Hyperacclerated, Through The Superconductive, Slush Hydrogen Plasma, Of A Toroid Cyclonic, Flywheel Ring - At An Infinite Expansion, Scale Of Magnitude: Tethered X Shelf, Continuity Positioning, Systemically Maintained, By Homeostatic Equilibrium, To Achieve Optimal, Power Factor, Universal Field, Unified Valence Output - In The Situation, Of Forming, Aging, Giant, Dwarfing, Or Colliding/Subsumed Stars, New Interdimensional, 4D Worm Hole, Temporal Displacement Events, Spontaneously Occur, Creating The Velocity, Energy Transfer, Of Dark Matter Distribution, To Maintain Spacial Acceleration/Rotation/Energized Matter, Field Integrity @ The Relativized Speed, Of Light, As It Extends, In Outward Trajectory, From The Big Bang Epicenter - Creates Doppler Shifts, Detected, By Large Optical Telescopes, Confirm Photon Particle Acceleration, From "Genesis Space" To Expanding, "Radiating New Space", Furthermore Determining, By Extrapolation, That We Are Currently, In The Middle, Of A Universal Field, Expansion Space Cycle - The Specialized Space Indicated, In The Form Of Worm Hole, Induction/Deduction, Quantum Sets, Indicate Flux And Flow Continuity, @ A "Little Bang" Magnitude, Much Like The Piston, Reciprocal Dynamics, Of A Perpetual Engine, Within Simultaneity, Sans Thermodynamic Entropy Breakdown, As It Occurs Outside, Of 3D Condition Timespace, In A 4D Cartesian, Non-Volumetric, Virtual Plane, "Undifferentiated Energy/Space" - Of Co-Existent Location Points, Like A Single Targeted Isotopic Particle, Existing, In Simultaneity, At Two Ends, Of A Flywheel Rotated Linear Accelerator, This Electro-Magnetic, Instantaneous "Bussing" Of 4D Introduced Matter, Although Degradated, In The Confines Of 3D Space, Is An Excited Virtual Particle/Anti-Particle Conduit, Which Helps To Produce Hyper Motility, And Velocity, Energetically Stepped Up, Paradigmatic, Power Factor Shifts, As Well As Spacio-Temporal, Quantum Leaps, In Adjoining, "Shared Space Interface" - Like The Di-Opposed, Microwave Actuated, Magnetic Bucking Coils, On A Mechanical Gyroscropic/Cyclonic, Propulsion Flywheel, Accelerating To Quantum Leap, Warp Speed, As Previously Indicated - The Induction Force, A 4D Phenomena, Increasing, In Tangent, With Mass Times Velocity Squared, Eventually Drawing The Hyperaccelerated Object, Into Itself, Once Sufficient, Forward Momentum Has Been Achieved, The Frictionless, Vacuum Of Space, Existing @ Absolute Zero, Of Non-Temperature, Exists, As A Superconductive, Plasma Medium, For Inducted Energy Transfer, Into The Velocity, Terminal Envelope, Of 3D Acceleration, Of The Propulsion, Quantum Dynamic Envelope - In 4D Induction Hyperspace, Quantum Simultaneity Sharing, The Same Properties, And Characteristics, Of A Double Hulled, Super-Refrigerated Vacuum, Of A Fusion Reactor, Tokamak Core, With Magnetically Trapped, Hull Contained Gamma Radiation, Fusion Reaction Offgassing, Serving As A Cold Fusion, Magnetic Bubble, Coolant/Insulator - Recreating, 3D/4D Vacuum Condition, Refrigerated, Superconductive Space - Magnetic, Ionic Trap/Bottle, Reactor Produced Anti-Matter Co-Valence - This Achieved, In The Reverse Manner, Of Linear Accelerators, Inducting, Dark Matter, Into Magnetic, Isolated Containment, Anti-Matter, "Farming Field", Like A Black Hole, White Hole = "Worm Hole Dynamic Set", Linear Accelerator Particle, "Induction Canon", Allowing **NO TIME**, To Elapse, For Anti-Particle

Anti-Matter, Transference, Within Introduced, 3D Condition Space, "Reacclerating Degradation", To Reaction Occur: In The Fusion Simultaneity, Of The Instantaneous, Anti-Particle Ionic/Dark Matter, Reactor Transfer, In Effect, Creating, A "Breeder Reactor" Array, To Replenish, The Depleted Fissile Material Required, Of The Toroid Ring, Becoming, A Replication, Of A Rotating Star, Of Ball Scalar Lightning, Inducted Slush Hydrogen, Cyclonically Energized, Into A Perpetual Engine, Warp Drive Propulsion System - These Are The Parallels Of Solar Physics, To Fusion Tokamaks

80) In The Alternative Case, Of The, "German Bell", Propulsion Is Achieved, Along The Lines, Of Both A Toroid, Ionic Hydrogen, Slush Ring, A Mechanical Flywheel, For Navigation, And Convention Rocket Propulsion, In That, The "Sky Iris" Induction Aperture, Not Only Inducts, Free Zero Point, Environmental Space Energy, But Also Through Core Reaction, Bombardment, Of Ionic Fallout, Creates Sidereal Pressure On The Interior, Reactor Bell Walls: The Sky Iris, Serves, As A, "Pressure Relief Port", For This Scalar Activity, Further Excited, By The Twin Adjustable, Tesla Induction Coil, Central Axis Inverted "Jacob's Ladder Array", Synergized, By A Magnetic Isolation Transformer, "Commutator Diaphragm", Which Serves, As A Super Efficient, Scalar Wave "Spark Gap, Between The Anode And Cathode, Termini, Of The Central Axis Point - The "Sky Iris" Relieves, Ionic Bombardment, Upon The Interior Facia, Of The Bell Walls: Thus Becomes, The "Jet Propulsion" Accelerator, Further Relieved, By The Anti-Gravity, Mag/Lev, Zero G, Properties, Of The Bell Base, Hydrogen Plasma Ring Continually Producing, Cyclonic Thrust, And Torc: The German Bell, Continuously Accelerates, In The Vacuum, Of Deep Space - The "Sky Iris" Aperture, Serves, As Mechanically Adjustable, "Jetting Port" To Regulate Both, Free Energy Induction Propulsion Rates, As Well As Controlled Release, Of Hyperaccelerated, Fallout Particles: Which Eventually Match, And Then Exceed, The Speed Of Light, As They Are Replaced, By Magnetically Teased, Induction, By Influx, Of Anti-Matter - During Black Hole, Tachyon Field, Toroid Field, Production - Reaction Occurring, At The Quantum Leap, Hyper Acceleration Range, As Indicated Previously - The Rugged, Metallurgical Construction, And Balanced, Weight Distribution, Of The Bell, With Ideal Resonant Cavity, Propulsion Characteristics, Add Increased Mass, To The Lorentz Invariant, Power Factor, "Mass" Compensating, For A Lack, Of "Load" Mechanical, Gyroscopic Parts - The Plasmoid, Slush Hydrogen Ring, Becomes A Chemo-Nuclear, Cyclonic Flywheel: The Mass, And Reaction Chamber, Provided By The, Bell Configuration, Hermetically Sealed, Double Contour Hulled, Provides A Pumping Reservoir, For The Stored, Slush Hydrogen: The Flushing Action, Of The Plasma, Increased, Proportionately, With Forward Acceleration Rates: Through Attending, Plasma/Fluid Dynamics - The "Sky Iris" At The Base, Of The Bell, Also Exposed, To Open Space, Acts As An Adjustable, "Thermostat", To The Reactor Core As Well, As The Slush Hydrogen Ring: Therefore, Serving A, Triple System Function: Essentially, Using Space Vacuum, Absolute Zero Temperatures, As A Hyper-Efficient, Radiator/Coolant, Of All The Components, Thus Dynamically Scaled, To Achieve The Minimum, Of Thermodynamic Breakdown, With The **LEAST** Amount, Of Moving Parts, In Vacuo, Rendered, **FRICTIONLESS**: Released "Balloon" With Escaping "Air"

81) The Next, Technological, "Paradigm Shift" In The Evolution, Of The Destiny, Of Man Requires, The Ultimate Goal, Of Achieving, Level One Civilization: The Ability To Protect The Planet, From Large Scale, Meteorite Strike/"Planet Killer" - Comet, Atmospheric

Entry/Superheating, Sky Defense, Against Incoming, “Hostile” Unidentified Flying Objects (UFO’s) Clean Non-Polluting, Super High Valence Energy Systems, Capable Of Inducting Zero Point, Ubiquitous/Ambient, Untapped Atmospheric, Inexhaustible Energy Reserves, Cataclysmic “Act Of God”, Weather Suppression, And Regulated Control, Repair And Cleansing, Of All The Atmospheric Layers, With Particular Focus, On The Thickening, Of The Ozone, Radiation Shield Layer, Grown Dangerously Thin, And Increasingly Ineffectual - Satellite, Theta Wave Emitting, Frequency Generation, To Attune And True, Planetary Rotational, Oscillation Rates, To Prevent, Earthquakes, And Tidal Waves, Decentralized, Levitational, “A/G Energy/Reactor Platforms”, That Can Be, Readily Dispatched, Into The Eyes, Of Hurricanes, Tornadoes, Active Volcanoes, To Subdue, Quell, Mollify Or Neutralize, Natural Disaster Outcomes, Through Free Energy Induction - Anti-Missile, Missile Defense, A/G Propulsion System, Aerospace Technology, For Viable Space Colonization, Low Cost, Free Energy Desalinization/ Purification, Of Oceanic Water, Turning Deserts, And Other Hostile, Earth Environments, Into Terra Formed, “Massive Land Reclamation Projects” - Theta Wave Frequency, Further Correlates, To Increased Intelligence, Reduced Violence In Human Populations, Therefore Reducing, To The Point Of Elimination, Globalized War Conflict, Along With The Attending, Mercantilistic Straining, On World Economies, Paradigm Shifted, To Research And Development, Of Generative And Productive, Aforeresaid, Level One Civilization, Enterprises, Essential In Continued Human Survival: With Off Planet, Colonized Proliferation, Of The Human Race: Fusion Energy Drive, Powered Tokamak, Propulsion Systems, Are Essential To Breach, Time Warp Limit Velocity, To Make Space Travel, And Exploration, Of “Earth Like” Planets, Technologically, Economically Feasible
82) The Grim Alternative, Of Malthusian/Draconian, Population Reduction, Will Simply, Temporarily, “Stave Off The Inevitable”, As So Many Other Nihilistic, “Final Solutions” Ironically “Culling Off” The Most Precious, Earth Resource, The Human Population - Effectively Assuring, Eventual Extinction, In The Long Run: By Thus Preventing, The Quantum Leap, Paradigm Shift, Into Level One Civilization, With Another, Dark Age In It’s Stead - Any Devolvement, Or Suppression, Of Technological Progress, As Human History, Has Graphically Demonstrated, Creates, A Backslide, Involving Hundreds Of Years: With Catastrophic Delays, In All Areas, Of Human Development At Current, World Population Levels, Many Magnitudes Greater, Than The Middle Ages - The, “Misery Index”, And The Ability, To Culturally Rebound, From Such A “Malthusian Scenario”, Would Be, “Proportionately Increased”, Perhaps Impossible, To Regain, Lost Ground, Due To The Scientific Complexity, Of The Current Industrial, Technocracy - The Progressive Solution, Seems Abundantly Clear, Ultimately, It Is An Categorical/Moral, Imperative Matter, Of Developing Zero Point, Free Energy, To Reach, A Level One Civilization, And From That Basic Juncture, Colonize Space - Additionally, Many Of The Rare Isotopes, For Such Required Advanced Technology, Will Be Found Abundant, On Other Planets, Occurring Naturally - Exploration/Exploitation, For These Vital Resources, Should Be, The Secondary Priority - In, A Level One Civilization, Evolutionary Paradigm Shift, “Futurist Civilizational Scenario” -
82) A Theta Wave Frequency, Wave Oscillation, “Earth Tuning Correction” Demonstrates Quantifiable, Biospheric, Ecological, Stability Enhancements By Observed, Empirical Extrapolation, Enhance Latent, Evolving, Neuropsychological, Abilities/Capabilities, Within The Ninety Percent Underutilized/Unused/Dormant,

Untapped, Portions, Of The Conscious And Unconscious Human Brain, Additionally Parapsychological/Metaphenomena, Would Also Increase, In The General Population, Enhanced In Particular, By Interfacing, With Theta Wave Driven, Biomechanical Engineering, Apparatus Required, To Navigate Warp Drive, Propelled Vehicles, In Hyperspace - Thus Developing And Accessing, New Neurological Human Potential

83) The Unconscious/Subconscious Mind, In Hypnogogic Trance/REM Sleep Stage/ Lucid Dreaming/Theta Wave, Deep Meditation State - No Longer Required, In The Case, Of REM, "Dream Stage Sleep", Specifically: To Regulate Body Motor Function, Which Has Been CNS, Temporarily Paralyzed, The Medulla Oblongata And, The Nerve Ganglion, Of The Neck Vertebrae, As Well As Endocrine Regulation Of The Hippocampus, Amygdala Nexus, Optical Nerve, Inverse Hemispheric Cross Juncture, As Well As The Auditory Organelles, The Autonomic Nervous System, No Longer Required, To Regulate, Waking Body Function, Beyond The Maintenance, Of Essential Renal, Cascading Biorythmic Function - Has Access To The Accumulated, Freed Neurological Energy Reserves, That Would Otherwise Be Directed, To Conscious Movement, Thought, Waking Biological Function - This Surplus, Quantum Amalgam, In Heightened Vibrational Amplitude, In Effect Resonates Beyond, The Physical Constraints Of The Sleeping Brain, Reduced Task Multi-Function: Energetically Bilocating, On A Quantum Level, Projected Brain Released, Expanded Consciousness, Along, A 4D Event Horizon, As Freed Energetic Quantum Mechanics, Of Theta Oscillating, Brain Waves, Become Energetically Sufficient, In Physically Unconditioned, Pure Energy State, Transcending/Sublimating, The Matter Of Brain, Into The Energy Valence, Of Disembodied Mind - Sufficient To Induce, Parallax Perspective, Of Future And Past Events, By Such Vibrational Energy, Acceleration, Past 3D Conditioned, Time Constraints - The Human Mind Becomes An, Energetically Disembodied/Disentangled, "Bio-Organic Time Machine" - While Retaining, Intact Sentient Consciousness, Of The Events, In Dreamtime, Confirmed, ESP Accurate, By Rigorous Empirical Research - Using Reliable, Known Historical, Fact Check Verifiers

84) Quantum Leap Navigation, From Such Telekenetic/Parapsychological, Time Traveling, Dream Consciousness - Can Be Amplified, And Integrated, Into Electro Encephalogram Interface, For Though Initiated, Navigation Systems - Designed For Hyperaccelearted Aircraft/Spacecraft - To "Dream Locate" Intended Destination Points

85) The Chrononaut Navigator, Will Effectively "Dream" The Destination Point Telekenetically, Guiding, The Warp Drive Trajectory, Telepathically - Aided, By A Piezo Electric, "Nexus Buoy" Repeater Emitting, In A Modular Frequency Interval, That Resonates, Throughout, The Time Line, Causal Event Horizon, Like A Span/Guy Wire, "String" Through The Worm Hole - The Great Pyramids, On Earth And Mars, Where, The Pyramids On Mars, Are Larger, To Compensate, For Mars Being Only 53 Percent, The Size Of Earth - Indicating, Geometrical, "Ratio Scaling", Of The Structures, To Work, In "Tandem": In "Dynamic Harmonic Concert", With Each Other - These Structures, Are Theorized, To Be Part, Of A "Signal Buoy System" That Led Back, To The Alpha Centuri, Tri-Star System, Our Nearest, Celestial Neighbor - Created Originally, For Hyperspace, Navigational Telemetry - Long Before, The Advent, Of Egyptian Proto-Civilization - Such Neuro-Cranial, Navigational Linked System, Interface, WithNuclear Fusion Warp Propulsion, Is Known, As A, **"Dream Chair"** Navigation Array

86) The Constant, Piezo Electric Pulse, Of The, Pyramidic Signal Buoy, Establishes

Oscillation Wave Flux, Which Serves, As Distance And Velocity Counter, Which Become Units, Of Space Measurement, Ie, "Mile Markers" - The Echolalia Interval, Is A Valence Constant, Calculable, Beacon Frequency, Within The Wormhole, Much Like A, Heartbeat Biorhythm - 4D Quantum, Vehicle Navigation Positioning, Established In This Way - Triple Linear Accelerator, Open Delta Frequency Array, Sending Nikohaben, (Hobbled Man) Triple, "Lock Step", Phased, Quantum/Synergy, Particle Waves, Provide, Sonic Line, Of "Blind" Navigational, "Dead Reckoning", As Well: Within The 4D, Dimensionless, Wormhole Matrix, Hyperspace, "Quantum Conduit"
87) Egyptian Mythology, Of The Great Sphinx, Gives Ample Clues, To This Ancient Chrononaut Technology, The Helmet Indicated, Is A Cranial Interface, Navigational Guidance System, The Stretched Out Body, In The Form, Of A Prone/Distended, Lion, Is The Form, Of A Human Being, Molecularly Stretched Out, In Hyperspace - The Pharonic, Throne Depictions, In Egyptian Art, Of The Upright, "Dream Chair" - The Jewelry Encrusted, "Beard" A Breathing Apparatus, For Life Support - The Egyptian Anck, A Dream Chair "Sonic Key" Held, In The Right Hand, To "Thread" The Nikohaben, Resonating Frequency, Through Dead Reckoning, In Hyperspace Worm Hole Teleportation - "The Riddle Of The Sphinx", What Crawls On Four Legs, Then Walks On Two, And Then, **HOBBLES,** On Three: A Cane, Or Shepherds Crook, Is The Third **"LEG"** These Are The **Three Stages Of Man**: The Child, The Adult, The Old, Broken, "Crippled" Man = Referencing, The "Nikohaben" (German) Broken Vav (Hebrew) Lamed **6** (Lamed=Hebrew=**30**) Lamed Vav (Hebrew) = The **36**th, Prophet, Is The **"Broken Vav", "Lamed Healer"** Triple Frequency, Which When Synergistically, Frequency Entangled, Can Stealth, And Break, The Speed Of Light Barrier, Through Linear Flywheel, Particle Acceleration - The **"Anck Key"** A, "Tuned Resonator" "Compass Needle", To Track, The Nikohaben Signal, "Thread", Through Hyperspace - These Technologies, Became, **Mystical Symbols**, Indicating Early Chrononaut Contact, With Prehistoric/Precivilization, Egyptian Aboriginals: "There Is Nothing New, Under The Sun" - As The Case In Point - The Punishment, Of Edipus, King Of Thebes, Questioned By The Sphinx: The Incesting, Regicide, Patricide, Made, "Blind And Crippled" = Yet Another, Mythological Reference, To "Nikohaben Frequency", As Edipus, Is Still Able: "To Navigate The Earth, Though Lamed And Blind" Clear Indications, Multiple Culture, Primitive Contact, With Advanced, Unexplainable Technology: As Mentioned Before, The Legend, Of Balder/Thor, Resurrected From, A **"Lamed, Indo-Aryan, Healer"** - With The Ability, To Induct Scalar Lightning, With A Swirling War Hammer/Cyclonic Flywheel: After The Hobbled Man/ "Nikohaben", Strikes His Staff, Upon The Ground: He Is, Temporally Displaced, Into The Norse Sky God, Of Thunder - Wing Helmeted, Like Hermes, & Egyptian Sphinx The "Shepherds Crook", Is In Itself, An Indication, Of An "Induction Coil", Expressed In The Toroid Center, Of An Open Delta Circuit, By A Curvilinear, "Induction Rod" Which Can Be Substituted, For A Helmholtz, Span Coil, Or Tesla's, "Jacob's Ladder" The Ideal Shape, For The "Shepherds Crook" Induction Hoop, Is A "Golden Section" Theoretical Conical Spiral, Of The Universe, From The Big Bang: Spinning Outward
89) The Inter-Dimension, Of 4D Space Conditioned Space, Beyond "Pi Point" Interface, Has Two Unique Characteristics, Beyond It's Light Barrier, Breached, Spacial Void It Has, Both Infinite Perspective, Holographic Characteristics, And Non-Dimensional, Omnidirectional, Paradox: Not Unlike, An "Unfathomable", Black Box Theatre,

Where Any, Tangent Point, Is A Staging/Platform, Of Seeming, Horizontality - It Can, Be Passed Through, In Virtual Simultaneity, Like A Film, Membrane Barrier, No Calculable Width, In No Time: Or Be Stasis Occupied, By Magnetically Insulated Biological, Or Inert Matter, In A Timeless Eternity, In A Dimension, Void Of Physics, Required, To Generate, 3D Time Space - This Permanent Condition, Conspicuously Absent Of, Mass Times Acceleration, Squared: Creates An Eternal, "Non-Time Oasis" Which Can Be Scientifically Exploited, To Extend, The Human Life Span, Indefinitely -
90) This "Time Arresting", Phenomena, Can Also Be Replicated, In And Around, Productive Tokamak, Fusion Reactor, Slush Hydrogen, Toroid Rings, That Substantially Slow, The Decreased Rate Progression, Of Time Elapse, Upon Influenced, Surrounding Matter, Arresting Rates Of Material Decay: By Spacio-Temporal Displacement, Caused, By Inducted Ambient, Anti-Matter: The Same Energy, That Exists, In The 4D Field: Non-Time, Non-Space, Non-Matter, Pure Energy, Hyperspace Dimension This Phenomena, Also Occurs, Around Solar Magnetic Fields, Where Time, Is Also Warped, By Hyper-Gravity Fields, Upon Relatively Small, Orbiting Objects, And Around Super Dense, Imploded Dwarf Stars, With Their Original, Magnetic Fields Intact, Which Are Precursors, To Black Holes: Warping Time, And Consuming Light
91) Time Anomalous Conditions, Of This Nature, Should Increase, All Forms Of Computing, And Data Processing: Including, The Human Brain/Mind, Thought Process, As Neurotransmission, And Synaptic Firing, Should Reach, Optimal Biological Efficiency Various, Dormant, Brain Function Capacities, Will Most Probably, Begin To Pronounce Themselves: Including ESP, And Telepathic Communication: This Would Be Consistent, With A Plethora, Of Environmental Concentrated, Free Energy, Being Induction, Tapped And Utilized, In The General Dynamic Proximics, Of The 4D Stargate: An Interdimensional, Perpetual, Free, Pure, Limitless Environment Derived, Energy Source
92) Additionally, Chronic Medical, Perhaps Psychological Conditions, Of An Organic, Degenerative Nature, Could Be Kept In Stasis: Instead, Of A Progressive, Terminal Prognosis - Live Human Births, In Such, "A Time Anomalous Zone", Would, In All Probability, Endow The Child, With A Zero Point, Time Referent, Ideal For Potential Chrononaut, Psychological Profiling - However, They May Not Mature, At Normal Rates, Of Development: And Would Subsequently, Have To Be Transported, To Normal 3D/1G, Conditioned Space, To Mature Normally - Yet Still Retaining, The Zero Point, "Psychological Referent", Ideal For Time Traveling, Human Chrononauts
93) The Philosophy, Of Time Travel, Is A Four Dimensionally, Conditioned Relationship, To The Forward Time Line Monitoring, Stewardship, And Course Correction, Away From Cataclysmic, Past, Present, Future, Time Events: With The Advanced Technology, Comes With It, The Ethical And Moral, Responsibility, With Well Informed Historical Hindsight, And Empirically Observed, Foresight, To Make Required Adjustments, With The Least Amount, Of Direct Intervention, Generally, Employing Human, Often Hybridized Surrogates, Using Available, Technological Resources, At Their Disposal, In Real Time: So As, To Not Disturb The Intended, Natural Course, Of Human Events - This Must Be, Also Closely Observed, For Subsequent Causal Outcomes, With

Repeated, Chrononaut "Visitation" Adjustments - To Assure The Best, And Most Stable, Long Term, Future Time Line = Strong Indications Suggest, This Is Already Taking Place
94) Human History Abounds, With Such Contacts, And They Are, Well Documented,

In The Chronicles, Of All Ancient, And Modern Cultures - "The Mzunga" Of Central Africa, Who Came Centuries, Before European Colonization, Warning Of The Imperialistic Threat, To The Continent - "The Pahana" Of The Western, United States, Who Also Warned, Native Americans, Of European Invasion: And Encouraged, Confederation And Allegiance, Among The Warring Native Tribes, "The Aristotelean Society" A Mysterious Group, Of "Renaissance Cartographers", Who Provided, Accurate Sea Charting Maps, To The Americas - Many Modern Advanced Technologies, Are Retro-Engineered, In This Manner - These Also Tend, To Have, No Known, Credited Inventor, Attributed To Them: Which Is The Tell Tale Sign, Of Chrononaut, "Intervention" The Information/Technology, In Question, Generally Thus Given, In An, "Incomplete Form", Requiring, "The Recipient", To "Decode" A Large Portion, Of The "Schematic" Or "Cartography" To Be Of Utilitarian Use: And To Make It, Technically, "Their Own" - The Opposite Situation, Has Also Arisen, Where Nearly Complete, "New Science" Is Further Helped Along, And "Problem Solved" Years Ahead, Of Predicted Schedule, If The New Technology, Is Already, At A Nearly Finished "Solution State", Where It Will "Inevitably" Be Solved, By Normal Human Means: In This, "Indirect Manner", Of "Partial Disclosure", And Accelerated, "Foregone Conclusion" Scenarios, Of "Advanced Problem Solving", Cultural And Technological, Quantum Leaps, Are Precociously Attained, Paradigm Shifts Facilitated, And The Future Time Line, Further Stabilized, In A Generative, Forward Trajectory - When Cartography Maps, And Technical Schematics, Are Too Complex, Or Scientifically, Too Advanced, In Nature - Telepathic Communication, Is Established With, "Direct Ancestors", Of The Chrononauts, Oftentimes, Becoming "Walk In" Hosts, This Generally Occurs, In Early Childhood, And Continues, Throughout Life - This "Dynamic Relationship", Based On, "Genetic Matching", Through Ancestry, It Is The Most, Efficient Method, Of Disseminating, Appropriate Advanced Technology: Through Autodidactic, Polymathic Means: Without Interfering, With The Forward Time Line, "Correction Path", Adjusted Course Trajectory: Via, Total System Designing, Master Inventors, "Walk Ins"

95) The Ultimate Goal, Is To Develop, The Necessary Components, Physical Plant, Generating And Distribution, Energy Systems, To Reach, "Technological Parity" - With, The Aforesaid, Chrononauts: Who Have Been, In The Process, Of Preparing Us, For This, "Nexus Parity Point", For Thousands Of Years: As We Are, In Essence Their Past, And They Are, Our Future : Upon The Zero Point, Future Event Horizon Line

96) The Paradigm Shift, Toward Level One Civilization, Will Be Accomplished, Through The Viability, Of Controlled Fusion Physics, And It's General Scientific, Technological, Medical, Transportation Applications, From Limitless, Zero Point, Free Energy Induction: Meeting Expanding, Energy Requirements, To Achieve Planetary, And Space Colonization, Sustainability - At Present Population, Density Demographics, We Have Precious Little Time, To Accomplish These Ends - To Ignore, The Blatantly Obvious Ramifications: To Suppress Technological Progress, Towards These Ends, Effectively Eliminates, The Quantum Leap, Of Human Evolutionary Potential: That The Paradigm Shift, Of Collective Consciousness, Entails - Without, The Large Scale Ability, To Explore And Homestead, Willing Earth Colonists, On Terraformed Planets, With Similar, Atmospheric Conditions: The Doubling, Or Even Tripling, Of Current Population, Will Implode, Causing, A Total Collapse, Into Anarchic Crisis, Of A New Dark Age - All The Civilizational Gains Made, Lost And Forgotten, In The Ensuing, World Chaos

97) Without Forward Planning, And Resourceful Application, Of Alternative

Ultra High Valence, Energy Systems, We Will Go The Way Of, "Easter Island", On A Planetary Scale - Scenarios, Involving, Any Malthusian Attrition, Of The Population, Are Ridiculously Nihilistic, And Morally Unethical - In Light Of The Fact, That The Technology, In Question, Exists, In It's Fundamental Form - Research And Development, A Fraction Of The Cost, Of The Alternative Attempts, "To Cull The Herd" - It Is Therefore, A Win Win, Strategy, In The Long Term, To Go Beyond, Current Rocket, Aerospace Technology, Beyond Carbon Fuel Propulsion Systems: To The Solar Physics, Of Fusion Technology, & Time Travel -

98) We Have Already Crossed, The Threshold, Into Nuclear Physics, With Mixed Results - Rather Than Abandon, What Has Been Accomplished, Thusfar, We Must Go Beyond, Fission Physics, An Intermediate Stage, Onward, To A Power Source, That Can Be Utilized, As A Propulsion System, To Break, The Speed Of Light - Owing, To The Tremendous Distances, Involved, In Deep Space Exploration - A Perpetual Engine Source, Is Required, That Is Self Reliant, And Can Regenerate, It's Required, Fuel Supply, By Onboard Breeder Generation - It Also, Has To Create, 1XG Force, Of Internal Gravity, To Maintain, The Bone Integrity, And Psychological Stability, Of Chrono-Astronauts - Traveling, Over Extensive Periods, In Deep Space - The Vehicles Themselves, Must Be Powerful, And Large Enough, To Create Sufficient Living Space: Easily Escaping, The Gravitational Planetary Pull, Through Increased, Thrust Velocity: Without Exerting, External Physics, Within The Space Vehicle -

99) These, Are All Physical Properties, Inherent, In Anti-Gravitational, Magnetic Levitation, Fusion Physics, Warp Drive, Propulsion Systems - As Described - In The Paradigm Shift, Of Future Mankind, This Will Be, The New, "Noah's Ark"

100) Paradigm Shift, Implies Radical Change, In Current Human Consciousness, Which By Extension, Brings About, The Required, Collective Will, To See Through, The Necessary Changes Made, Into A Viable Means, Of Generative Production - Given, The Opportunity, And Aware, Of The Potential, Dire Consequences, Of Apathetic Inaction: Many Would Rise, To The Occasion, To See The Mission Through - Human History, Has Been, A Series, Of These Awakenings, And Evolutions, Of Expanded Consciousness: To Meet The Increasing Pressure, Of New Technological Challenges - That In Their, Ingenious Solution, Have Brought About, Quantum Leaps, In Civilization - Without Which, We Would Not, Be Here Today: Pondering The Next Paradigm Shift Stage, Of Our Next Technological Quantum Leap, Into The Undiscovered Realm, Of The Previously Unknown - Fortunately, As We Found, The World, Was Not Flat, As We First Circumnavigated The Earth - Nature Has Always Provided, The Perfect Model: Solar Physics, Is The Perfect, Perpetual Energy System: Meant To Sustain, For Billions Of Years - The Vacuum Of Space, An Energy Accumulating Plasma, Which Can Be, Induction Tapped, Using Gyroscopic Mechanical, Or Slush Plasma Toroids - It Is, A Case, Of "Technological Mimesis", To Synthesize, Solar Energetic Function, To Our Advantage Thus Attaining, To Level One Civilization, Required Technological, Goal Parameters

101) The Concept Of Time Itself, Is A Theoretical Abstraction, Derived From Physically Amalgamated, Environmental Dynamics, Operating Upon A Three Dimensional Plane - That Constellate, And Align, Electromagnetically, Into A Perceptible, Psychological, And Physical Phenomena: Exerting Influence, Upon Biology And Matter, Through The Life Cycle, And Through, Thermodynamic Breakdown Of

Technological Matter, Deterioration, And Transmutation Of Isotopes, And So Forth Any One Of These Variables: Matter, Times Velocity, Squared: Augmented Or Supplanted, In Their Synergistic, "Time Manufacturing" Dynamic - Will Cause "Time Displacement", Special Events, To Occur, Approaching 4D, A/G, Phenomena Removal Of All Energetic Components, Of The Stated Equation, Becomes The Pure Condition, Of 4D Space, Which Removes, The "Contingent Notion Of Time", As Physically Based, Temporal Reality, Is Removed, In Energetic 4D Hyperspace

102) Manipulating, Any Component, Of This Dynamic Triad, Underlining Causal Factors Can Facilitate, An Opportunistic, "Breach Point", In The Fabric Of Time Space, Producing Their Total Negation, In The 4D Plane, "Space Tear" - Revealing The Anti-Matter, Zero Point, Centralized Field Source, Where They Derive Inducted, Expendable Energy, Driving Themselves, In 3D Motility, Through Expanding Space

103) Any Form, Of Heightened Acceleration, Increases Output Valence Yield, And Systemic Stability - Therefore All Forms, Of Thermodynamic Instability, Are Caused By "Entrophic Slowing", Of Optimal, Rotational Characteristics, Of "Momentum Acceleration" - In Direct Relation, To Quantum Of Matter, It is Exerting Motive, Or Cyclonic, Force Upon - The Ideal Rotation, Or Motile Traction, Of Acceleration, The Axiomatic, Paradigm Profile, Emulates, Perfectly Trued, And Perfectly Balanced, Symmetrical Mass, Solar Rotation - In Graduated Scale, And Proportionate, Output Valence, Power Factor, Positive Production, Nuclear Fusion, Yield Rates -

104) All Other, "Special Events" In "String Theory", Reconciliations, With "Relativity", Are Exceptional And Unique, Chaos Entanglements, Are "Order Rectified", Once Sufficient Induction, Of Free Energy Is Reintroduced, To Restore Homeostatic Order - In The Temporarily Imbalanced, Physically Anomalous Event - Generally, These Chaotic Weaker Fields, And Amorphic Plasmatic Fields, Either Rearticulate, And Stabilize Themselves, Or Are Opportunistically Absorbed, Into Proximal Stronger More Stable Fields, Into A Higher Valence Amalgam, Combined Energy Gestalt -

105) Paradigm Shifting, Within The Human Mind, Also Based On Weak And Strong Field, Psychodynamics, Which Create, Morphogenic Resonance, "Empirical Set Replications" When Discovered, Beyond A Certain, "Compression Point", Of Dissemination: To Meaningfully Describe, New Discoveries, Within Scientific, Or Cultural Reality - They, Therefore Supplant, Outworn, Archaic Notions, Of Said Realities, In More Precise, And Accurate Ways - Thus Creating, The Milieu, For Paradigm Shift, Quantum Leaping, In Expanding Awareness, Of Consciousness

106) Technologically Enhanced, Human Progress, Towards The Inevitability, Of Progressive, Paradigms Shift: Through Axiomatic, New Conceptions, Will Create, A Cascading, Domino Effect, Of Geometrically Accelerated, Expanded Consciousness: In Part, Synergistically Amalgamated, By Unlimited Access, To Previously, Archive Accumulated, Empirically Vetted, Reinvestigated Old/New Science: That Cross Correlates, And Fact Amplifies, The Unified Field Theory, Of Fusion Simultanaety - The Scattered Pieces, Of A Long Pondered Puzzle, Of Ontology And Eschatology, Corresponding, To The Energetic Genesis, And Forward Momentum, Of The Nature Of Things: Whether, In The Microcosmos, Or Macrocosmos, Of Intentioned, Teleological Purpose, Of "Intelligent Design": Synthesized, Technologically, To Emulate And Harness, Solar Fusion Physics: Thereby, Achieving Level One Civilization: Through Zero Point, Power Factor Induction, Of Free, Unlimited Energy:

With That Limitless Energy Platform, In Controlled, And Perfected Form: Create Propulsion Systems, And Generator Platforms, Breeder Reactors, That Can Break The Speed Of Light: Instantaneously, Crossing Vast Sectors, Of Space, In Quantum Leaps, Through A Networked System, Of Wormhole Paired, Positioned Stargates - Allowing For, Unhindered Space Exploration, And Subsequent, Human Colonization Of Terraformed, Newly Created Worlds, Of Our Own Making - The Myth, Of Paradise Regained, The Return To The Garden Of Eden, No Longer An Impossibility - The Paradigm Shift, Also A Fundamental Change In Human Psychology: Freeing Us From The Limiting Confines, Of Previous Held, Archaic Belief Systems, Achieving Energetic Liberated Mind, Over Corporeal Matter: The Untapped Recesses, Of Consciousness, Awakened, From Vestigial Slumber: Fusion Technology, Dynamically Driven, By Theta Wave Frequency: In Turn, Brain Amplifies, Our Own Capacity, To Increased Intelligence: Concentration, Parapsychological Capacitance: Required To Navigate, Such Mind Controlled, Mind Operated, Fusion Propulsion Technology - The Human Mind, Was Created, For This Purpose, And Has Yet, To Reach, It's Full Biological Potential - Examples Of Greatest Genius, Through History, Attest To The Potential Range, Of Human Intellectual Endeavor - Instead, Of Being, An Anomalous Phenomena, Found, In A Small Percentile, Of The Population: It Will Become, The General Gestalt, Gold Standard, Of Psychological Functioning, In The New Paradigm Shift - In Effect, Causing, A Permanent, Evolutionary Leap, Forward, In Human Development: Fusion Technology, The Didactic Instrument, That Makes It So - "Mind Over Matter", Demonstrated, On A Physical Level, By Brain Driven Energy

107) The Astrophysical Dynamics, Of Blackhole Suns, And Their White Hole, 4D Twin Valence Set, Paired Counterparts: Are Antimatter Producing, Fusion Reactors, When Energy, Overcomes Matter, In The Compression Process, Of Dwarf Stars, Within Intact, Gravitational Fields, That Once Contained, And Concentrated, The Solar Physics, Of Hydrogen Chain, Perpetual Nuclear Reactions: The Remaining Negative Space, Once Occupied, By The Solar Mass, Now A Super Magnetized Vacuum, Serves, As A Tachyon, Antiparticle, Antimatter Generator - Inducting 4D Free Energy, To The Central Locus Point, Of The Remnants, Of The Hyper Dense, Supercompressed, Material Remnants, Of The Imploding, Extinguished Star - This Tear In Space, Sucks Surrounding Matter, As Well As Photon Light Particles Towards, An Open 4D Center, Like A Magnetic Vacuum, @ Acceleration Rates Beyond, The Speed Of Light - At The Same Time, Antimatter, Antiparticles, Dark Matter, Are Being Drawn Through, In The 3D Opposite Direction, Escaping, 4D Non-Space, Through The Magnetic, Black Hole, Interdimensional Tear - This Matter/ Antimatter, Exchange Point, Is The Basis, Of Creative Materialistic Energetic Force, Universal Propulsion Momentum, Plasma Acceleration, For Power Distribution, And Magnetic Material Consolidation, Of Newly Created Space - The Physical Distance Between Matched, White Hole, Black Hole Sets, Is Rendered Irrelevant: As Their Dynamic, Synergistic Valence Output, Collapses Time Space, Within The Worm Hole Conduit Set, Which Forms Between Them, Which Is Unconditioned, By Time - Two Ends, Of The Space Tear, Within The Magnetic Field, Acting, As An Antimatter "Containment Bottle" Is Linked, At Both Ends, Of The Space Tear, In Simultaneity Effectively Sharing Energetic Fusion, Denuded, Of Time And Space, Variability - Thus Synergistically Operating, In Their Twinned, General Dynamics: Like A

Flywheel Rotated Linear Accelerator, On A Macrocosmic, Cosmological Level
108) This Natural Cosmological Phenomena, Which Is Ubiquitous, Through The 3D/4D Interdimensional Universe, Is The Energy Network, By Which The "Big Bang" Effect, Is Perpetually Maintained, This Energy Exchange, Like The Steam Traction Piston, Of A Vertical, Flanged Flywheel Set, Of An Antique Locomotive, Their Track, Proximics, Based Upon, The "Roman Chariot Wheel, Gauge", Creating Environmental, Ambient, Rotational Energy Induction, The Motive, Locomotion Mass, Of The Vehicle, In Increasing Acceleration, All Contribute, To A Dynamic Envelope, Of Kinetic Energy, That Truncates Space Time, Through Mass, Times Acceleration, Squared - Taking, The Mechanical Elements Involved, And Reducing Them To Hyper Efficient, Nuclear Fusion, Energized/Actuated, Components, Augments And Amplifies, The Induction Profile, Of Free Energy, To An Exponential Geometric Scale Level - To Paradigm Shift Gradient, Where Free Energy Valence, Supercedes Accelerating, Amalagamated, Algorythm Physics, Of The Mechanical, Or Plasmatic System, Devise: Causing The Said, Hyper-Rotating System, Of "Locomotive Hyper Propulsion", To Be Temporarily Replaced, By An Energetic Field Replication, = "Valence Doppleganger" - Spacio-Temporal, Displacement Event, And Teleported Through, 4D Interdimension, To, A Matched Valence, Conditioned Field, In 3D Time Space, Where It Slows, Into Sub Light Frequency, From Warp Speed, And Rematerializes, Into Original Form:
109) Controlled Fusion Amplitude, Fusion Plasma Toroid Ring Sets, Creating Black Hole Star Gates, Can Accomplish, These 4D Quantum, Teleportation Quantum Leaps, In A Precise, Calculated, And Materially/Energetically, Uniform Way - Making A Regulated, Strargate System, To Desired Regions, Of Navigable Space: Impossible To Currently Reach, At Sublight, With Rocket Propelled Technology - Only By Replicating, The Cosmological Structure, And Fusion Physics, Of Black Hole, Magnetic Gravity, Of Imploded Stars - Reached, With Anti-Gravity Propulsion, And Magnetic Shielding Array - Can, Hyper Propelled Objects, Be Successfully, "Valence Converted", From Matter, Into Energy/Anti-Matter, And Be Successfully Teleported In Simultaneity, From Two Distant Synthetic, Toroid Induced, "Starpoints" Instantaneously - Any Other Propulsion Method, Would Simply, Take To Much Time, Requiring Generations, Of Astronauts To Reach - Consuming/Burning, Requiring, Unavailable Quantities, Of Combustable, Rocket Fuel, In The, "Rocket Propulsion Process": Regenerating Fusion Nuclear Reaction, The Way
110) Additionally, Without, A Gravitationally Controlled, Interior Environment, With 1XG, Of Earth Gravity, Human Bone, And Muscle Tissue, Would Begin, To Disintegrate, Without Gravitational Resistance, Stress And Counterforce: Within A, Short, Period Of Time, Soon Crippling, The Manned Space Crew - Making Extended Space Voyage, Physiologically Unviable: Psi Pressure, Of Brain, Spinal, Endocrine Fluids, Would, Hydrodynamically, Pressure Fluctuate, Creating 0XG Weightless, Psychotic Mental Illness - Thus, Technically, And Biologically Impossible, To Withstand, Or Life Sustain: In A Prolonged, Zero Gravity, Weightless Environment - Without Total Encapsulating, Magnetic Hull, Exterior Radiation Shielding, Facilitated By Fusion Warp Power Generation, Space Radiation, Solar Bombardment, Would Lethally Penetrate, The Interior Hull Cabin, In The Nuclear Radiation Spectrum - Undeflected Extreme Heat, Which Is Not, Magnetically Repulsed, Would Also Prove, A Certain Death Sentence, Unabated, By The Open Vacuum Conditions, Of Deep Space, No Filtering Atmosphere, To Block, Or Deflect, Their Frequency Penetrating, High Capacity,

Upon Unshielded, Vacuum Exposed Matter - Creating, An Interior Microwave Oven
111) Therefore, Fusion Reactor Based, Magnetic Flywheel, Magnetized Plasma Toroid Induction, Based Upon, An Electro-Magnetic, Power Platforms, Producing Gravity Fields, And Magnetic Bubble Hull Shielding, "Reciprocal" Breeder Induction Reactors, Is The Only, Logical Propulsion Source/System, For Future, ANY Deep Space Travel - Involving Manned, Exploration, Or Mass Colonization Missions -
112) In The Paradigm Shift, To Come, All Will Be Made Well - As Chronicled Patterns Of Human History Clearly Indicate, Paradigmatic Shifts, Are Closely Linked, To Millennium Periods, And Aeon Transitions - Where The Most Anxiety Driven, Mass Hysteria, Tyranny Of The Masses, The Tribal Apotheosis, Of Archaic Pessimism, Along Apocalyptic Doomsday Themes: Retrogressive Superstition, Countered In Hive Mind, Cult Mass Consciousness, By The Most Unique, Creative, Original Thought, In Antithetical Postulation, Gleaned From The Most Individuated Genius Minds, Existing Intellectually, AboveThe Frey - Many Cultural Traditions, Ancient Religious Belief Systems, Have Insightfully Speculated, This Backsliding Trend, Peculiar Devolution, In Consciousness, At Strategic, Historical Axiomatic Points - As A Virological Etiology, Targeting, The Human Reptilian Brain, And The Organelle Confluence, Of The Cerebrospinal Fluid - Which May Be, Of Extraterrestrial Origin, Conveyed, By Particle Beam Bombardment - In Ancient Times, This Was Attributed To Sirius, Turning Red, Associated With, Great Epizootic Outbreaks, Among Warm Blooded Mammals, Working Itself Up The Food Chain, Eventually Effecting Mankind
113) These Strange Anecdotes, Recorded By Many, Including Homer, In The Illiad - Were Generally Known, Portents, Associated, With A, Climatological Raise, In Atmospheric Temperature, Draught, Plague Conditions, Associated With Sirius Turning Red, In Late August, In Twelve Year Cycles: Known As, The "Dog Days" Of Summer, As Sirius (=DOG) And Dogs, Were Often, Initially, Negatively Affected - This Pattern, Was Discerned, By Philosophers, In Various Unconnected Cultures - As An, "Intentioned Design", To Keep, Human Intellectual Development, Arrested - At Key, Transition Points, And Paradigm Shift, Nexus Points - With The Ultimate Aim, Through Time, Of Keeping Human, Technological Progress, Below "Level One Civilization", And Away From Nuclear Fusion, Propulsion Capability: Marooning Us, In The Long Term, On An Overcrowded, "Prison Planet" Like Dumbed Down Cattle: Collective, "Hive Mind" Cult Consciousness, Seems To Attend, Hysteria Stressor, Symptomatology, Of Millennialist Periods: Psychological Profiles, Seem To Conform To This Hypothesis: The Attending Mythology, Indicates, An Ancient, Reptiloid Race
114) It Would Seem, Regardless, Of Past Notions, @ The Current Nexus, Of Human Technological, And Psychical/Psychological, Development, Has Reached, Parity, With The Alleged, "Reptilian Perpetrators", Indeed, Perhaps Surpassing, This Legendary, Adversarial Source: Attributed, To The Historic Downfall, Of Mankind - No Longer, It Would Seem, Are Fully Capable, To Manipulate, Large Demographics, Of The World Population, As Was, The Case, In More Primitive, Ancient Times -
115) The Real Possibility, Of Exo-Biological, Life Forms, As Well As, Encountering Advanced, Alien Civilizations, Seems Highly Probable, As Space Exploration Becomes Commonplace: This Will, Bring About, Yet Other, Paradigm Shifts, Etc.
116) The Current Rise, In World Climate Temperature, Caused By Greenhouse Gasses, And, The Combustion, Of Fossil Fuel: Along With The Deforestation, Of Massive

Oxygen Producing, Equatorial Land Tracks, Including The Flora And Fauna, That Is Being Brought, To Extinction: Is Raising The Mean Temperature, By Ten Degrees Over Time, And Altering The Biochemistry, Of The Earths Atmosphere - Essentially Altering And Terraforming, The Earth: Potentially, For A "Reptiloid" Cold Blooded, Life Form - Conspiracy Theory, Abounds In This Area - Suffice It To Say - Level One Civilization, Protocols, Of Climate Control, Including Seasonal Regulation, Of Earth Temperature, Increasing The Density, Of The Ozone Layer, And Preventing Solar Radiation Waves, To Effect Wind And Ocean, Increased Weather Instability: A NATURAL "Sky Defense", That Has To Be Maintained, And Cleansed Of Industrial Pollutants, That Have Been Proven, To Deplete And Thin, As Well As Make Toxic, The Water, And Air, As Well As, The Food Grown, From These Contaminated Sources - Nuclear Fusion Reactors, Are Capable, Of The Extreme Energy Levels, To Desalinize, Rising Oceanic Water Levels, Derive Hydrogen Fuel, From Hydrolysis: A Combustable Gas, The Returns, To Water Vapor, And Is Non-Polluting As Well As, Satellite Repair, And Cleanse, The Upper Strata, Of Earth Atmosphere, Of Detected, Chemical Contaminates, Through A Program, Of "Electromagnetic Sluicing" Of Bonded, And Neutralized, Toxic Compounds - Huge Amounts, Of Generated, Ocean Desalinized Water, Can Be Reintroduced, To Recover Desert, And, Draught Wasteland: The Increased Greening, Of The Earth's Surface, Will Naturally Restore, Optimal Oxygen Levels - And Regulate Rising, Atmospheric Heat Once, This "Restorative Technology" Is Perfected, It Can Be, "Space Exported", To Terraforming, Of Other, Earth Like Planets - Well Before, The Human Colonization Process, Has Begun - Using Automated Equipment, Sent Through Hyperspace - This Also Represents, A "New Economic Model", The Application, Of Limitless, Zero Point Energy, To Planetary Restoration Projects, & Advanced Space Technology - And Away, From Obsolete, Carbon Fuel Based, Petrochemical "Toxic Industries"
117) Of Man, Myth, And Legend: Of Prometheus Unbound, And Atlas Shrugged - While, Nephilim, Anunnaki: By Archimedean Leverage, In Paradigm Shift, Reversal: To Halcyon, Golden Age, Utopia Polity, Mind Projected, Into Righteous Futurity - Human Liberation, Through Quantum Evolution, Enfolding Beyond, The Arcane, Laws Of Destiny: When The Artificial Barrier, Of Time, Has Been Left, Behind Us: No Longer, A Penitentiary Wall: Technologically Transcended - History, No Longer Repeating Itself, Like, A Children's, Puppet Show - The Pessimism, Of Existential Neurosis, Eternal Return: Hegemony Lie, Imposed, Upon The Slave, Relevant, No Longer: Unfathomed Fate, Negated, Whole Truth Be Told: W/Discerning Awareness Vetted Prophesy, Holy Testament, From Time Immemorial: Man Long Oppressed, Held Back, In Raging Fury: Like Thoughtless, Slaughter Warring Chattel: Once Contemptuous, To The Leviathan Beast, Seal Of Baphomet: Man-Counterfeiting, Viper, Unable To Dissimulate, Illusion Upon Us: They Shall Beware, The Discerning Time, Of Judgement, Cometh Nigh - When They, Are Plainly Seen, With, Mark Of The Beast, Upon Them - They Cannot Withstand, Earth Colony Exodus, Nor Follow Gathered Humanity, Beyond The Wall Of Time - Relegated, To The Pit Of Hell, For A Thousand Years: Curse Condemned, Upon Serpentine Belly, Under Human Heel Their Shapeshifting, Vile Form Revealed, Reptiloid, Souless Eyes, Snake Blinking: ("The first Chinese Emperors, successors to the legendary, Emperor Yao, said they were, descended from dragons. In India, high caste families still believe that their

ancestors, are snake gods called "Nagas." The Dogons, of Mali, claim that their extraordinary, astronomical knowledge, came to them: from their, Ancestral Reptilian Celestial Aspect, the "Nommos." The Egyptians, had their god snake "Kneph" and the Pharaohs were often depicted with snakes. The Phoenicians had "Agathodaimon", a serpent. Cultures of Central America, worshiped their creator god, "Quetzalcoatl", the winged serpent, as "Baholinkonga", the feathered serpent, of the Hopi Indians. Followers of Voodoo, have a serpent god, they call,"Damballah Wedo". The Hebrews, have "Nakhustan", the Brazen Serpent. The ancient god, of the British Isles, called "Hu", was the Master of the Dragon World. The first altar, of the Cyclops, in Athens was dedicated to "Ops", the God-Snake. Indigenous Australians, also have their myth, of "Wagyl", a God-Serpent.") The Ancient Greek **"Stigma"** Is Expressed, By A "Snake Pictogram", Numerical Value = **6**, Hebrew, Letter/Number, When "Broken" (Broken Vav) Is Used To Describe, A **"Serpents Belly"** = Strategically Positioned, @ The "Belly/Center" Of The Torah Scroll = **6**
"Eyes Wide Shut" = The Reptiloid Iris, Blinking Sideways, **"With Eyelids Open"**
118) A Clear Pictures, Emerges Here, From The Oldest World Cultures, In Both Eastern And Western Civilization: Of A Serpentine, "Dominator Race" That Hybridized, Into The "Ancient Aristocracies", Of Said Civilizations: Which Became, More Advanced, Technologically, While At The Same Time, More Repressive, Of The "Oppressed" Homo Sapien, "Mass Host Population": While The "Snake God Religions", In Question, "Flourished", They Where, Historically Noted For, Unparalleled, Ritual Human Sacrifice, On A Holocaustic Scale - Instead Of Expanding, After Reaching, A Savage, "Pinnacle Point", Of "Barbaric Sophistication" - After Which, The "Snake God's", Mysteriously Disappeared, Without A Trace" The Civilizations Collapsed, And Rapidly Disintegrated, "Snake Worshipping, Aboriginals" Devolving Rapidly, Back, Into Their "Original, Autocthonic State", Indicating, A Synthetic And False, Temporary, Unoriginal Human Progress: To Serve The "Blood Lust Agenda" Of The Encroaching, **Exobiological Life Form**: Upon A Primitive, And Easily Manipulated, Mass Human Population - In Other Words, **Anti-Civilization**, The Clearly Defined, Oxymoronic Antithesis, To The Quantum Human Evolutionary Goal, Of Level One Civilization - The Nag Hammadi, "Gnostic Gospels", Further Fill, "Historical Gaps" Of Unexpurgated Scripture From 200 BC, Alluding To Similar **Reptilian Incursions**
119) By Extension, "Level One Human Civilization", Is A Resounding Answer, To This, Historical Challenge - Which Is, Technologically Superior, Nuclear Fusion Energy Based, And Involves, Human, Off World Colonization: In Paradigm, And Population Shift, From, A Previous, "Parasite/Host, Victimology Stance", Ending Of "X Factor", Dominator/Malthusian, Nihilistic, "Human Population Culling", "Old(Ancient)/New", (Modern) = "Strategies" To Cultural, And Civilizational, Level, Of Self Determination
120) As Opposed, To Remaining Terrestrially, "Marooned" While Consuming, Our Limited, Remaining Resources, As We Overpopulate, The Planet, To The Point, Of Total Exhaustion/Depletion, Deforestation, Etc - Fossil Fuel Combustion, Toxic Destruction, Of The Atmosphere, Through Greenhouse Gas, Global Warming, Solar Irradiation, From Ozone Depletion = Genetic Mutation, Food Depletion, To World Famine Levels, Population Explosion, Related Opportune Plague Conditions Associated, With Demographic Overcrowding - Climatological Conditions, However "Well Suited", Raised Globally, By Ten Degrees, For, **Reptilian Life Forms,** To

"Thrive" - **Ancient History, Repeating Itself?** With Similar, Human Catastrophic Results? - One Has To Question, The Insane Rationale Of, **"Business As Usual"**
121) In Summation, The Goal Of Unified Field Physics, Is Primarily, To Rectify Environmental, Global Toxic Conditions: Caused, By Several Hundred Years, Of Unchecked Corporatized Industrialism, Based On Carbon Fuel Economy/Inefficient, Costly, Polluting, Energy Generation - Causing, A Progressive, "Greenhouse Effect", Which Is Now Reaching, Obvious Critical Mass, Worldwide Impact, And Proportion - Particularly Deleterious, To The Radiation Blocking/Shielding, Depleted And Thinned Ozone Layer: Accelerating, The World Mean Temperature, Through Endothermic Heat Induction, Of The Ozone, Unprotected Atmosphere, By A Staggering, Ten Degrees, Fahrenheit - Enough To Create Major Latitudinal, Ecological, Environmental Shifts, Towards, The Equatorial, Temperature Peak, Which May Prove, Irreversible, @ Continued Carbon Fuel, Combustion Rates - As The Second, And Third World Nations, Begin To Industrialize, En Masse - Exacerbating, Pollution Ton Yields, Thousandfold -
123) Clean Forms, Of Nuclear Cold Fusion, Zero Point Energy Induction Platforms: As Well As, Other Non-Polluting, Energy Systems, Currently In Place - Can Further Help Terraform, Otherwise, Uninhabitable Earth Environments, Desalinize Ocean Water, To Potable Levels, Of Drinking Grade Purity, Provide Reservoir Supply, For Land Irrigation, And Work, Dynamically/Synergistically, Within Power And Generation, Current Power Grid Distribution, On A Wireless Transmission, Satellite Based, Globalized Level -
124) Increased, Power Factor Output Valence, And Zero Point Induction, Amalgams, Means Cheaper, And Exponentially More Powerful, Free Energy/Nuclear Fusion, Energy Supply, In Vast Surplus, Of Required Demand Curve - Critical, In The Further Development, Of Power Engineering: Required, For The RanD, Of Advanced, New Science, And Cutting Edge, Medical Advancement - Everything, From Toroid Frequency Yokes, Implanted Around, Viable Optic Nerves, To Cure Blindness, Through Cybernetic Videographic, Photon Transmission, Bypassing Or Amplifying, Damaged Eye Efficiency To Similar, Fusion Powered, Nanotechnology, To Synthesize Cures, For Various Degenerative, Sensorium Organelle, Array Deficits - High Energy Power Plants, Sufficient To Drive, Exoskeletal Prothesis, Allowing Debilitated Amputees, To Use Previously Damaged, Or Have Replaced, Biomechanical, Functional Limbs, Able To Walk Again - The High Energy Required, For Such Complex, Mechanical Devices, Reduced, To A Small Battery Sized, Power Module -
125) Intelligent, Computer Guided Nanotec, Actuated, By Micropropulsion Systems, Supercharged, By The Same, Miniturized, High Energy Power Plants - Downsized To The Point, Of Injectable Form: Can Both Site Locate, And Perform Non Invasive, Lazar Amplified, Pin Point Targeted, Nano-Robotic Surgery, Through The Circulatory System, With Ultra Precise, Disease Targeting - As Well As, Sending Diagnostics, In Real Time - Able To Monitor, The Patient, In the Post Surgical, Healing Phase, 24/7
126) Electro Magnetic Imaging, Can Be Made So Powerful, As To Be, On Parr, With An Electron Microscope, Or Better: So As To, Detect Anomalies, On A Subatomic, Or Genetic Level: Detecting Any Range, Of Biological Abnormalities, And Location Correcting Them, In Real time, With Automated, "Micron Level", Surgical Precision - With Miniaturized, Linear Accelerators, Serving As, Computer Controlled, Beam Output, Amplified "Lazar Canon", "Micro-Scalpels": Such Medical Advancement, Will Be Necessary, In Exploration, Of Deep Space, Safeguarding Against: "Exo-Contamination"

127) No Longer A, "Ship Of Fools" - Lost At Sea, Reduced, Through Increased, Survival Needs, And Population Compression, To The "Anarchic Piracy", Of "Competitive Struggle", Into, Zombified, "Flying Dutchmen" = Marooned, As We Are, On This Prison Planet, Earth, Turned Toxic, "Penal Colony", Over Time, As World Population Explodes - Beyond The Planets, Resource Capacity, To Sustain Itself - We Must Therefore Fix, What Has Been Broken, While There Is Still Time, To Reverse, What Has Been Damaged - From The Same, Advanced Technology, Create Space Stations, Stargate Networks, To Initiate The Process, Of "Mass Exo-Colonization", No Longer A Thing Of Science Fiction, Speculation: But A Looming Reality, Of Continued Human Survival -
128) As In All Things Revolutionary, There Will Be Much Resistance - But Also Many Willing Participants, Who Will, Readily Volunteer, In The Grand Experiment, The Ultimate Odyssey, The Greatest Adventure, Mankind Has Ever Known - Like The Halcyon Golden Age, Of The Heroes - Many Will Rise, To the Occasion, Others Will Be Born,To Undertake, The Herculean Task, Set Before Them - When The Self Limiting Constraints, Of Pseudo-Science, In The Superstitious Form, Of Modern False Religion, Is Rigorously, "Empirically Challenged" - And The "Big Lie", Hegemony, Of Ivory Tower, Half Truth, Is Rationally Overcome - When New, Theoretical Paradigms, Align With The Ultimate Reality, Of "Astrophysical Phenomena" - Fathoming, Through, "Technological Mimesis" And Thus Replicating, **The Unified Field, Meta-Phenomena**, Of Nuclear Fusion, Solar Physics -
128) Only Then, Like "Prometheus Unbound", From Caucus Mountain, Slavish Servitude, To Cruel And Unusual, Wrathful Olympian Gods - Will We Realize, Our Condition, Of Obstructed And Frustrated, Manifest Destiny: And Thus Embrace, Our Full Evolutionary, Human Potential, As Was Intended - Manned, With Self Created, Genius, Tools Of Ascent - Like, "Jacob's Ladder" = Capable, Of Interdimensional Travel, And The Discernment, Of New, Bold Discovery, Once Thought Unimaginable -
129) The Last Concept, When Talking About, Temporal Displacement Technology, Is, The "Philosophy Of Time Travel" One Must Consider Well, Such Things - We Are NOT Alone, In This; There Are Those, Who Have, "Watched Us", From The Beginning - And Become Our Stewards, Our "Guardian Angels" Over Time: Their, "Prophets Mantle" Will One Day, Fall To Us: As They Too Evolve, To Their Ascending Destiny, Such Is The Way Of Things - As We Serve In Their Place, When The Time Has Come - As We Serve The Highest Power - To Guard Our Future, As Well As Our Past, Our Ancestors And Our Descendants, Upon The Event Horizon Line - Like The, **"Hippocratic Oath"**: A) Do No Harm B) Relieve Pain And Suffering C) Inform And Advise, But Do Not Interfere D) Leave **NO** Advanced Technology, Physically Behind, Temporally Disturbing, The Natural, Evolutionary, Time Line E) Unless It Be, In Schematic, Or Cartographic Form, In An, "Incomplete State", Requiring Solution, In Real Time F) Approach Only, Your Own Direct Ancestry, Ancestors And Descendants, Who Are Thus, Intuitively Aware Of You G) Only They, Will Understand, Your Righteous Purpose, Recognize, And Willingly Help You H) Personality/Identity, Fusion, "Walk In's" Can Only Be Accomplished, Through Direct Ancestry, For The Psychological Bonding Process, To Be, "Twin Mind" Accepted
130) Ultimately, What Will Be, Will Be - There Are, Outer Limits, To "Forward Time Correction" Certain Historical Events, Are Found To Be, Unchangeable, Therefore Intrinsic, To The Fabric Of Time, And Are, "Meant To Be" - By Their Nature, They Are

NOT, To Be Tampered With, Or "Event Manipulated" - Being Of, And From, The Will Of The Creator, Therefore Inviolate - By Following, These Basic Protocols, And "Failsafe Precautions", The Philosophy, Of Time Travel, Is Therefore, "Ethically Maintained" -
131) Human Tissue Regeneration, In Lieu, Of Organ Transplantation, Can Be Excited Into Accelerated, Cloning Production, To Repair And Replace, Damaged Body, And Organ Systems - Using Genetic Mitochondrial, Human Genome Information, As Master Schematic - Inducted Fusion Energy, Serves As A Zero Point, Power Factor, "Incubator" To Accomplish, This Tissue And Organ, Regeneration Process: Through Experimental, Trial And Error, The Properly Attuned, Valence Oscillation, Band Width, Can Be Isolated: That Safely Accelerates, Growth Of Living Biology - Creating Viable, Transplantable, Matched, Genetic Tissue, Replication - Eliminating, The Need, For Organ Donation
132) In Such Matters, Of Such Time Visitation, Atemporal Events, Use Informed Judgement: Approach Only Those, Who Discern And Fathom, Your Nature, Do Not Overstate, But Guide: What Is Already Ascertained, By Their Own Knowledge - Approach Ancestors, And Descendants, Who Easily Recognize, Your Family Resemblance, To Them: Minimizing The Potential Risk, For Traumatic Shock - In This Way, Minimal Time Incursion, Is Experienced, In The, Unified Fusion Field -
133) Telepathic, "Psychological Adjustment", Of "Would Be", "Notorious Individuals" Circumvents The Need, For Physical Altercation - Which Will Effect Generations, Of Their Descendants, If Lethal Force Is Used - This, Only Employed, As A, "Last Resort" - With Extreme Prejudice, When All Other Methods, Of "Forward Time Correction", Have Been, Completely Exhausted - And Their Effect, Upon The Time Line, Considered To Be An Unacceptable Level, Of Foreseen Risk, To The Forward Time Correction, Outcome -
134) Previously Unsuccessful, Time Experiments, Conducted With Sufficient, But Unstable, Valence Output, On The Earths Surface - Cause Temporal, Time Gap Flux, Tears, That "Reaction Echo", In Approximately, Thirty Year, Reverberating Cycles, Generally Taking, Two Such Cycles, To "Self Correct" - Emulating, The "Rotation Cycle", Of Saturn's, "Lunar Year" - This Is Caused, By Primitive Conduction, Commutation Methods, And Improper Acceleration, Deceleration, Methods, In The Reaction - Which Must Be Symmetrical And Instantaneous - "Reaction Asymmetry", Conversely, Creating, Localized, "Devil's Triangles" The Antithesis, Of A Productive, Controlled, "Open Delta": The Asymmetrical Reaction, Creating White Hole/Black Hole, Tears, In Localized Space, Oftentimes, Quantum Leaping, Thousands Of Miles, In An Uncontrolled, Unified Field Arc = Obviously, Superconductive Plasmas, Or Mechanically Regulated, Flywheels, Are Required, To Establish Controlled Rate, Of Ionic, Isotopic Rotation, In A Stabilized Field
135) Symmetrical/"Unilateral, Fusion Reaction, Simultaneity", Within, The "Dead Short", Of An, "Open Delta Circuit", To The Centralized, Induction Toroid Core: Elementally Comprised, Of A Triangulated, Direct Current (DC) Power Source, An Opposing, Alternating Current, (AC) Power Supply, Of Equal Power Factor, And A Helmholtz Coil, Span Wire, W/Flux Capacitor, Frequency Oscillating/Zero Point Energy, "Induction/ Broadcasting Tower" With A Span Wire, "Free Energy, Influx Antenna", Equal In Length, To The Total Height, Of The Two, Connected/Supporting, Flanking Span Towers - Placed Equidistant, In A Triangulated "Open Delta Formation" = Electromagnetically Commutated, To Each Other, And Simultaneously, "Electronically Bussed" To The, "Centralized, Super Magnetized Toroid" In The Center, Of The "Open Delta, Triangle Configuration" - Will Create, A Nuclear Fusion Reaction - When An Object, Is Placed

Within The Toroid, Temporal Displacement, "Central Ring", Itself Equipped, With An Antagonistic, Magnetically Di-Opposed, "Internal Field" = "Teaser Array" @ Equivalent Valence Output, To It's Own, Gross Atomic Mass, Further Voltage Ballasted, And Electronically Equipped, With Asynchronized, Tuning Thyristor, To Further Excite, The "Internal Mass Reaction", The Net Effect, Will Be, The Temporal Displacement, Of The Said Object, Into 4D Hyperspace: As Long As, "Current Production", Is Symmetrically Distributed, From All Triangulated Points, Are In Simultaneity: Within The Nuclear Fusion, Energized, Open Delta Array - Once The Trifold, "Nikohaben" Frequency, Generated Energy, Is Symmetrically Terminated: The Toroid Encapsulated, "Independently Magnetized"="Target Object", Will Reappear, In 3D Conditioned Space, In It's Original, Solidified Form, Unchanged: Down To The Molecular Level, Of Structural Integrity - The Internal, And External Magnetic Fields, Must Be Proportionately, Electronically "Mag/Lev Isolated", Equally Repulsive, And "Magnetically Antagonistic", To Assure Reaction Stability, Within, The Unified Fusion Field, Which MUST Occur, In Electronic Simultaneity = These Are, The Advanced Technology Elements, Of Teleportation, & Viable Fusion Reaction Powered, Temporal Displacement/Time Travel
136) Given The Necessary Perimeters, For A Symmetrical Temporal Displacement Reaction, Employing A Matched Set, Of Slush Hydrogen Energized Rings, Equally Energized, By A Tokamak Fusion Reactor Valence, Power Distribution Array - Would Be, The Implementation, Of A Magnetically Levitated, Monorail Tracked "Bullet Train" "Sled" Anti-Gravity, "Propulsion Actuated", Then Stepped, Up In Hyper Velocity, Through A Large Scale, Linear Accelerator "Tunnel" Which Would Sent The "Sled Vehicle" @ Sufficient Acceleration, Through The Twin Set, Toroid Rings - The "Time Capsule Sled" Should Theoretically, "Disappear" In Between, The Tracked Distance, Between The Ring Sets, Appearing Instantaneously, Externally In 3D Space, On The Other Side Of The Outside Opposing Ring - Acceleration Can Be Eliminated, By Cutting Of The Power Supply To The "Mag Lev, Traction System", Of The "Time Capsule Sled" Or Reversing Magnetic Polarity, Of The The "Mag Lev Train Truck" Reversing It's Function, Into A Powerful, "Track Gripping Magnet", Effectively Becoming, "A Clutch Break", On The Monorail, "Right Of Way" - The Trackage Would Have To Be Of Sufficient Length, To Allow, For Thyristor Controlled, De-acceleration, Electronically Monitored Crash Test Dummies, Filled With "Cadaver Synthetic Gel" Can Study Internal Gravitation, And Temporal Physics, On Potential Human, Chrononaut Subjects, Within A Magnetically Gaussian, "Faraday Cage", Internally Integrated Fuselage, Equipped With Life Support, And Autonomous Gravity Field - After Technological Refinements, It Can Then Be Safely Adapted, To Advanced Aeronautical Engineering, As Well, As Fusion Physics, Technological Applications, Currently Employing Nuclear Fission Reactors, And Jet Turbine/Solid Thiokol, Rocket Fuel Propulsion, Energy And Propulsion, Power Generation, And Exospheric, Earth Escape Velocity, "Gravitationally Contingent", Rocketry: Converted To, "Magnetically Enveloped", Heat/Radiation Shielded, Deep Space Ballistic Platforms, With Perpetual Fusion Reactor, Acceleration Power Plants -
137) There Is Nothing New Under The Sun, Especially, In Relation, To The Solar Fusion Physics, Of The Sun: Whether It Be, The Botanical Configuration, Of A Plant, Or The Piezo Electrically Activated, Solar/Free Energy Induction, Of Egyptian Pyramids, Step Pyramids, Babylonian Ziggurats, Gravitationally, Mass Stress Pressurized, Malachite

Limestone Temples/Walls, Mezo-American Pyramids, Etc, They All Function, On Some Degree, As Primitive, Low Valence, "Stargates", When, Solar Radiation Bombarded - By Their Immense, Gravitation Load, Which Inevitable Compresses Them, Into Concave Geometric Form, Along Lagrangian, Syntopic Plane: Piezo Electrically Generated, Magnetic Frequency Lines, They Take On, Solar Characteristics, Drawing In, Environmental Free Energy: The Issue, Is To Take These, "Fundamental Elements", Involved, In These Ancient, "Holy Geometry" Meta/Physics, And Reduce Their "Generative Properties", To Pure Fusion Physics Elements, Of An Operational, "Induction Generator", That Can Then Defeat, Gravitation Pull, While Productively Creating, More Energy, Than It Utilizes: In Geometric Proportion, And Gradience Of Magnitude, In Hyper-Efficiency: Creating, A Power Factor, Great Enough, To Ultimately Overcome, It's Own Specific Mass: Beyond Mere, Anti-Gravity, "Lift Surplus", Into The Realm, Of Temporal-Spacial Displacement, Warping Events - It Is In, This Ultra-Efficient State, & Electro-Mechanical, Manner, Utilizing, Nuclear Fusion Reaction, Power Source: That Synthesized, Solar Physics, In All Phases, Of Observed Astronomical Phenomena: From Convening, Gaseous Plasma, To Loosely Particulated, Glowing Nebula, To Coalescing Mature Star, To Expanding, Red Giant, To Gravitationally Imploded, Dwarf Star, To Black Hole Sun: Becoming In Effect, A Time Anomalous, "Solar Magnetic Star Gate", It's Own, Atmospheric Gravitional Pressure, Creating, A Super Dense, Matter/ Anti-Matter, Rotating Magnetic, Induction Core: Can Thus Be Emulated, As A Propulsion And Perpetual, Free Energy Inducing, Ultimate Power System, Fundamentally Consistent, With The Most Powerful, Natural Energy Source, Currently Known To Science - The True, "Promethean Fire" Once Given To Man, Is Nuclear Fusion In Nature
138) "The Open Delta", Free Energy, Induction Circuit, Is A Fascinating, "Electronic Paradox", Seemingly, Wired In Triangulation, Between An (AC) Alternating Current/ Oscillation Wave, Generative Power Source, A (DC) Direct Current, Flat Band Width Frequency, And A Helmholtz Coil: Curvilinear Span Wire, Flanking Twin Tower, Symmetrical, "Power Modulation", Open Delta Circuit, "Ballast/Controller", All Centrally Shunting, Simultaneously, Disparate, Energy Wave Forms, Into A Central High Energy, Magnetized Toroid Ring, In Apparently, A, **DEAD ELECTRICAL SHORT:** All This Accomplished, Without Shorting Out, Or Electrically Destroying, The Equilaterally Triangulated, Peripheral, Power Generating Apparatus, To The "Centralized Targeting Toroid Core" = Part, Of The "Paradox Explained", Is The "Tensile Deflection", Of The "Arch Tuned" Helmholtz Coil, Span Wire, "Suspension Adjusted", To A "Tangent Of Offset", Which Inversely Corresponds, A "Neutralizing Counterbalance", To The Earth's Gravitational Pull: Like A Electro-Magnetized, Microwave, Exciter/Inductor, Frequency "Harp" - The Central, Magnetized, Toroid Coil, Wired Into A Dead Short, When Super Energized, Draws Upon Inducted, Environmental Free Energy, Into It's Open, Ring Core, Effectively Creating, It's Own, "Magnetically Contained", "Independent Electronic Field", Which Also Serves, As A, Super Efficient/Valence "Stepped Up" = "Magnetic Isolation Transformer", Thus Disallowing, A Short Circuit, "Earth Grounding" In The System: Due To, The "Insulating Encasement", Of The Central Toroids, Energy Induced, "Magnetic Bottle" Within, The Open Delta Circuit: To Maintain, "Electronic Continuity", Without, Electronic Shortage: It Must Be Turned, On And Off, In Precise Simultaneity - Any "Target Object", Within The, Tachyon Field, Of The "Central Ring Toroid", Must Also Have, A Competing, Antagonistic, Repulsive, Magnetic Field, To Be Successfully,

Spacio-Temporally “Displaced” - Successfully, In A Symmetrically Stable, Curvilinear “Yoked” Tensor, Isometric, Interdimensional Field = (ie.,Inter-Dimensionally, Teleported) , Energetically Proportionate, To It’s Effected, Gross Gravitational Mass, Times (X) Velocity (V) (= Triangulated Shunting, Ionic Energy/Frequency, “Rate Flow”) Squared (X2)=Synergized, Into A Triple Wave, “Nikohaben”, Multi-Oscillation, Energizing/ Informing Current: In Order, To “Quantum Leap”, Beyond The Conditioned, Atmospheric, Photon, Light Barrier - The “Inducted Surplus”, Of Available, Free Energy, Garnered, “Outside” The Simultaneous Fusion Reaction, Energy Array, Amalgamated, “Productive/ Generative, Output Valence”, Also Competes, Repulses, Excites, Antagonizes, And Ultimately, In So Doing, Magnetically Insulates, The Internal Toroid Ring, From Being Destroyed, By Incoming, Triangulated Wave Form, Modulated Shunted Energy, In The Form, Of A, “Dead Short”, By Creating, A “Magnetically Teased” Energy Backwash, Against The Introduced Power Flow: Causing, An Interdimensional, Phase Shift “Warping”, As The “Generated”, And “Inducted”, Energy Fields, Dynamically “Collide” @ The Center, Of The Open Delta, Where, The Electronic Short, Would Typically Occur: Were It Not, For The “Magnetically, Insulating Properties” Of Ultra High, “Power Factor Induction”, Of The Surrounding, “Free Energy”, In And Around, The Power Generating, Electronic Shunting, Distribution, Component System, Within The, “Triangulated Apparatus”, Of The Instantaneously Energized, Dynamic Insulated “Open Delta Circuit”
138) Such Artificially, Magnetically Synthesized, Tensor Core, Non-Collapsing, Hyper Spin Rotated, Isometrically Stabilized, Lorentzian, “Twin Wormholes Sets” = (Scharzschild, Einstein-Rosen Bridges = Eternal “Traversable” Black Holes) As Demonstrated, By Gyroscopically Spun, Fusion Nuclear, Particle Accelerators: In A Casimir, Simultaneity Effect, Of Anti-Matter/Dark Matter, Induction/General Quantum Field Stabilization - A Negative Mass, Cosmic Ring Set, Influx Exciter, Free Energy “Venturi”: Mimetic, On A Microscopic, “Localized”, Torn Space, “Throat” Field Range, To Energy Matter Exchange, “Particle/Anti-Particle, Shifting/Inter-Dimensional, Free Energy Oscillation, Which Occurred, In The Big Bang: And Continues Ubiquitously, In The General, Infinitely Expanding, Universal Field Valence, Space Vacuum, Plasma Physics Dynamic, Eternally - Superluminal, Time Dilation, Determining Wormhole, Spacio-Temporal Metrics: For Feasible, Distance Placement, Of Black Hole, Nuclear Fusion Energized, Open Delta, Ring Toroid Sets, Would Therefore, Technically Allow, For Intergalactic, Truncating, Time Travel, And Feasible Scientific/Logistic, Methodology: Necessary, For Mass Human Colonization, Of Habitable, Terra Formed Planets - The Primary End Goal, Of Level One Civilization: Exo-Habitation, Of Suitable New Worlds
139) The Cyclonically Entwined, Energy Whirlpool, Inter-Dimensional, Stargate Portal/ Conflux, At The Peak Valence Epicenter, Between The Toroid, White/Black Hole, Space Tear Sets - Is Kept From, “Tunnel Collapsing”, By Axis Leveraged, Tensor Points, @ The Vortex/Conflux, Of The Dipolar Archimedean, Conical Screw Like, Spin Rotation Thyristor, “Regulated” Nuclear Fusion, Slush Toroids = Through, Whirlpooling, Particle Spin, And Ambient Free Energy, Scalar Induction = Oscillation Modulating, Relativistic Space Time Confluence, Into An Anti-Matter, Magnetic Bubble: In Astrophysical Consistency, With A Fully Collapsed/Imploded, And Ultra-Gravitationally, Compressed Black Star, In Vacuo - The Thermodynamic Contrast, Of Exothermic, Rotational Gradient Scaling, Towards, The Spacial Nexus Tear - Further, Thermally Insulating/ Isolating, And Stabilizing It, From Spacio-Temporally, Manipulated, And Arc Fused,

Interdimensional, Linear Vortex Euclidean, And Curvilinear Lagrangean, Astrophysical Tensor, Magnetic Lines, Of Rotation Spun, Cyclonic Force: Heading Towards, The Central Stargate, Space Tear - Establishing Symmetry, From Opposing Directions
140) Incoming, Surrounding, Hyper-Accelerated, Tensor Field Spun, Whirpooling Particles, Matter, Velocity Induced Free Energy - "Fuel", The Twin Toroid, Paired Reaction, In Simultaneity: In Symmetrical, Mass/Velocity, Amalgamated Power Factor, From Opposing Termini = Like, A Blast Furnace Bellows, Providing Increased Oxygen Supply, To A Forging Foundry - The Nuclear Fusion Reactor, Twin Quantum Toroid, "Time Machine" Converting, Oncoming/Incoming Matter, From Opposing Directions, Of The Compounded, "Space Cones" = Worm Holes - Into Dimensionless, Anti-Matter, At The "Breach Yoked" Center, Twin Scalar Cone, Stargate Nexus: Thus, Informing, The Non-Physical, Non-Spacial, Exotic Characteristics, Of The Central Yoke Tear: Wormhole Piercing, Non-Dimensionally/Interdimensionally: Through, Folded Space Time, Between Intervening Terminus, Tangent Points - Circumventing, Surrounding Distance And Time: Through, The Dimensionless, "Yoked Tensor Ring", Scalar Space Tear: Shortcutting Between, The Surrounding Vacuum, Of Conventional Space Time, Like A Hot Knife, Through Butter: Abrogating And Truncating, To Zero Point Distance/Duration, In General Field Simultaneity: In Scientific Theory, Actually Existing, In Both Places, At Once = The General, Unified, Physical Quantum Mechanics, Generated, By An Interfacing, Cyclonic Valence Set, Of Nuclear Fusion Toroid , Of Synthetic, Or Naturally Occurring, Dark Matter Generated, Blackholes - Bisecting, At At A Zero Point, Time Tunnel "Yoked", Interdimensional, Stargate Portal, Which Have Equilateral "Righting" Spin/Valence Yield Tendencies, Towards Homeostatic, Perpetual/"Eternal" "Perfect Rotation", As The Magnetic Pull Influence, Of One Nuclear Fusion, Activated Toroid, Corrects And Spin Modulates, The Entwined Other, And Vis Versa: This Power Factor, "Twin Rotation Tensor Field", Scalar Wave, "Commutated Continuity", Is Absolutely Essential, In Stabilizing, Frequency Modulation, Of Super-Exited Particles, Converted, Into Anti-Matter, Returning, Through The Other Side, Of The Time Tunnel, "Isthmus Stargate", In 3D Rematerialized, In Symmetrical Simultaneity, W/Original Molecular/Genetic/Genome, Biological/Structural, Pre-Entry "Integrity": Only Time And Space Being Reaction Altered This Paradigmatic Axiomatic Parallelism, Of Entwined Helix Simultaneity, Is Time Travel: The Unified Field, Of Nuclear Fusion Physics, Equilaterally Distributed, Power Factor, Which Bonds, Accelerates And Expands The Universe, At Superluminal Velocity, "Magnetically Towing" All Amalgamated Matter, In Known, Contiguous General Space, @ 10, To The 94th Power @ Grams, Per CM, Cubed = Power, Times Velocity, Squared
141) The Fuselage Double Hull, Of The Space/Time Vehicle, Can Be, A Tuning Thyristor Regulated, Faraday Caged, Fusion Energy Powered System, Of Telsa Coil Construction: Further Supporting, A Linear Accelerator Array, Generated "Magnetic Bubble Envelope" - Which Can, Through Spark Gap, Resonant Induction Degaussing, Repulsion Deflect, And Pulse Flux Capacitor, "Bussing" Discharge: Create A Scalar Wave Induction, Input Valence "Trap", To Energetically Neutralize, Exterior Hull Radiation Bombardment: General Field, "Shunting", Lethal Levels, Of Radioactive Frequencies, And "Fast" Ionic Particles, Away And Around, The Magnetically Shielded Hull, From Ever Entering, The Interior Environment, Of The Capsule - Thus Neutralizing, The Charged, "Nebula Like Effects", Of Cyclonically Agitated, Oscillation Wave Excited, Exterior Radioactive, Space Bandwidths: Intrinsic/Characteristic Of: The Van Allen Belt,

Major Solar Flare, Eruptions/Activity, Irradiated Plasma Cloud, "Arcing Proximity", Encountered, Within Deep Space, During Extended, Space Missions - As Well As, "Magnetically Insulating", Biological Material, During, A Controlled/Stabilized, Wormhole Quantum Leap, Through "Two Point Vectoring" Simultaneity, In Energized Hyperspace -
142) Synchronized, And Counter-Synchronized, "Valence Parity" Toroidal Matched Pair Sets, Hyper-Rotated Into "Cyclonic Cone" Lagrangian Magnetic, "Caducean Modulation" @ Intersecting, Twin Cone Pinnacle, Inter-Dimensional Nexus, Creating, A Fusion "Tethered Linkage", Anti-Matter, Stabilized Vortex, "Tensor Yoke", Serving As A Energetic, Omnidirectional, Nuclear Fusion "Focusing Coil" - In The Epicenter, Of The "Reaction Matrix" - Whose "Juxtaposition, Tensor Fugue" May Determine, Spacio-Temporal, Exotic/Unique, Fluctuations/Variants, Backwards And Forwards, Upon The Linear Elapse, Time Causality, Of The Conventional, Event Horizon Line - As Well As Temporal Displacement, In Dimensionless Teleportation, Of Hyper-Energized, Effected Matter: Rendered In Simultaneity, Between Predetermined, Spacially Calculated, Twin Vertices, As "Target Vectors", Of Rematerialization, Back Into, 3D Conditioned Space: "Tele-Projected" White/Black Hole, Exit Points, Of The Twin Valence, Toroidal Termini -
143) The Free Energy Induction, Tensor Spin, Whether, Uniform Throughout, Within The "Entwined Cyclonic, Cone Pairing", Antimatter Yoked Configuration, Or Counter-Rotated, In A Competing, Antagonistic, "Repulsion Field", Would Either, Accelerate, Space Time, Truncated And Catapulted, In Quantum Leap, Into The Future Event Horizon, Or Reverse It Chronologically: While Still Maintaining, The Fixed Euclidean Integrity, Of Calculated Vortex, Terminal Target Points: Irregardless, Of An Influencing Time Factor, External Element, "Variable" - Fixed Spacial Tangents, Would Thus Act, As "Reaction Constants" - Both In Physical Positioning, And The Accelerated Matter Transporting, From Dimensionless "Phantom Energy Bilocation" Into Stabilized/Re-Organized, Original, Material/Biological Form: "Existing" Simultaneously, At Both Terminal Ends, Of The Teleporting Reaction, @ Once - This, Is Absolutely Imperative!
144) Any Slowing, Or Dampening, Of The Warp Reaction, Would Invariably, Alter Or Mutate, Or "Chaotically Transform" The Rematerialized Transported Object - Regardless Of Timeline Trajectory, To Future Or Past: Therefore The, "Time Factor", Within The Reaction Itself, Must Unilaterally Conform, To Zero Point Simultaneity, Throughout -
145) A Teleportation Event, That Is Not Occurring, In Zero Point Synchronized Simultaneity @ Both Termini Ends, Of The Matched Toroidal Set, Would Chaotically Scramble, And "Arbitrarily Reconstitute", Molecular Structure And Composition: As Parts Of The, "Unstable Worm Hole" Projected Object, Would Both, Accelerate Into Dimensionless Energy, And Also, Decelerate, Into Virtual Oscillating, Phantom Matter, @ Uneven And Unstable, Discontinuity Conversion Rates, Over The General Mass -
146) The More Unstable, Uneven, Chaotic And Asymmetrical, The Abortive Time Reaction, The Greater Effect, It Will Have Environmentally, Upon Surrounding Normal Space - Requiring, A "Forward Time Correction", To Help Resynchronize, The Spacio-Temporal, Calibration, Upon The Event Time Horizon: In The Twin Vectors Proxemics, Of Effected Space - Creating, A "Bermuda Triangle Effect" Of Destabilized Time Space -
147) This Uncorrected Time Anomaly, Occurs In Calculable, Deintensifying Cycles, Of Diminishing Occurrence: Eventually Reorganizing, From A "Chaos Conditioned Field" - Eventually Readjusting, Into "Causality Homeostasis", Within, A "Failed Reaction Event", Readjusting Algorithmically: Like Fall Out Rates, Of A Nuclear Detonation - However,

The "Reactor Source" Must Be, Energetically Eliminated, For This, "Readjusting Period", To Spontaneously Occur - Engineered Failsafes, Should Be, Design Built, For These Anomalous "Flux Outcomes" Until The Technology In Question Is Reasonable Perfected
148) The Primordial Law Of Nature, "Invest In What Grows", The Adaptive Radiation, Of What Evolves And Changes, In Accordance, With The Adjusting Needs, Of Empirical Reality - The Antithetical Apotheosis, Of Archaic Nihilism, In Superstitious, Fundamentalist, Retrogression: Devolves Eventually, Into Cowardly, Civilizational Decline, And Ultimate Extinction: Technological Progress, Is Thus Fundamental, To The Continuance, And Proliferation, Of Life Itself - To Understand, And Mimetically Synthesize, The Unified Field Fusion Physics, Of Ubiquitous Nature, Which We Are, Part And Parcel, Of - Which We Have Endeavored, To Gradually Discern, Since The Philosophical, Dialectical Speculations, Of Sage Antiquity: It Is, The Unique Intellectual Capacity, Of The Human Mind, To Quest, For Such Elusive Knowledge, Wisdom And Understanding: To Fathom Our Condition, And Dynamic Relationship, To The Universe, Which We Exist In: Not Merely, As A Passive, Malingering And Slothful Spectators, Contemplating, From A Disengaged, Safe Distance - Nor In Ignorant Apathy, Letting Time And Opportunity, To Render Progress, Surpass Us, As The Fossilized Record, Of Our Culture Stultifies: Beyond Our Window Of Opportunity, To Quantum Leap Forward, Into Our, Self Potentiated Future, Elapses Into Dust - As The World Was Found, Not To Be Flat, Nor Surrounded, By Monstrous Sentinels, Of The Deep - Convenient, Limiting Mythologies, To Keep Us, In A Fearful Mental State, Of Petrified Annui, And Unenlightened, Dark Age - So Too, Like Those Bold, Intrepid Explorers, Who Circumnavigated, The Oceans Of The Globe, Against The Flawed Wisdom, Of "Prejudiced Certitudes" - Discovering New Continents, And Civilizations, In Their Tireless, Exploring Wake - We Now Live, In Such A Scientific, Paradigm Shift, Where The Means Of Production, For Deep Space Travel, And Exploration, Are Within Our Technological Grasp: Similar Debates Rage, As They Did Then, Between, The Naysayers And Protagonists, Of Innovation And Exploration: Those Who Hold Fast, To Archaic Notions, Foundation Stones, Of Their Academic, Ivory Towers: Holding On, To Outworn Concepts, Once Considered, Original Speculations, Turning Approximated, Pseudo-Science, Of The Bygone, Past Ages, Into Technological, False Religion: Obstructing Modern, Research, Creativity And Innovation - Suppressing Anything, That Competes, With Their Massaged "Facts", And "Scientific Absolute Law" Groomed, Vain Sophistry - Serving Only, In The End, To Boldly Impassion, The New Generation, To Reach Their Own, Radical, Unique Conclusions, Empirically Verified, Through The Reality Of Demonstrable Fact, Of Advanced Discoveries, Make Their Own Indelible Mark, Conquer Space, Matter And Time - Scientifically Mastering, The Unified Field, Of Fusion Physics, Thus Reaching, The Paradigm Shift, Of Level One Civilization -
150) Technological Advances In Metallurgy, Finally, Must Be, Meticulously Pondered: To Be Industrially, Aerospace, Design/Built, Considered, And Manufacture Implemented, For A Deep Space, Intergalactic, Interdimensional, Space Warping Vehicle, With Sufficient Fuselage, Exterior/Interior, Support Structure Durability: The Composite, Alloyed "Super Material", In Question, Must Have Stable, Density Integrity, Exceeding The Transuranium, Periodic Table Atomic Sequelae - @ Or Above A 115 Baseline, As a Stabilized, "Bitching"/Hardening, Material Catalyst, For Property Density, That Is Impervious, To Inbound Radiation Bombardment, In The Gamma Ray Spectrum, Solar

Flare Range - As Well As The Light Warping, Tachyon Particles, That Constellate, Around Black Hole Induction, Antimatter Tensor Events, With The Unique Capability, Of Penetrating All Known Heavy Metals, And Displacing/Distorting, Biological/Molecular Structure, Including Lead And Graphite, Shielded Containment: Traditionally Employed, To Transport Fissile Material -

141) Post Uranium Weighted, Periodic Elements, Fused With Calcium, Atomic Weigh 40, Infused With Gold 78: Crucible Fused, Molecular Union, With A Fusion Nuclear Reactor Crucible, Magnetically/Rotationally, Tensor Spun, Into A Stabilized, Molecular, And Isotopic Excitability, Super Alloy, Hyperconductivity, With Stealthing And Cloaking Capabilities, Through Electro-Magnetic, Molecular Hyper-Acceleration, Beyond The Visible Light Spectrum - Will Further Promote And Engender, Phase Shift Quantum Of Said, "Super Material" To Hyperspace Energy, Transition/Conversion, As A Superconductive, "Supermolecule Medium" - While Keeping The Life Supported, Modular Interior, Hermetically Sealed, From Deep Space, Radioactive Bombardment - Additionally, Such A Structurally Complex Molecule, Will Have "Memory Form" Characteristics, Repulsing Structural Stress, And Perhaps Programmable, To Change Shape, Through CAD/CNS Like, Advanced Computer Technologies, Hard Wired To The Fuselage, Capable Of Morphing, When Nuclear Fusion Energized - Machining And Tooling Of Such Material, Would Have To Be Accomplished, In A Magnetically Sealed, Robotic Assembly Plant, Due To The Extreme Temperatures, High Magnetism, Radioactive Fallout - Cutting And Welding, Of "Supermaterial" Components, Annealing Of The Hull, Cutting And Installation, Of Structural Elements, Would Require Fusion Powered, Plasma Torches, To Maintain Molecular Consistency, Throughout The Entire Structure - Ultra Pressure Testing, To Check For Any Structural Cracks, Microscopic Flaws, Or Holes, Would Be The Final Physical Text, Exerted On The Interior And Interior

142) The Capsules, "Magnetic Shielding", Interior Hull "Faraday Cage" De-Guassing Would Additionally Act, As A "Double Failsafe", Exterior Hull, Radiation Bombardment Protection, Efficiently Prefiltering, More Easily Deflectable, Low Valence Penetration, Lethal Penetration Frequency Band Widths Of Space Radiation - Acting As An Artificial Magnetic Atmosphere, Around The Vehicle, Against The Unprotected Vacuum, Of Deep Plasma Irradiated Space -

143) Such A Super Alloy Cast Hull, Would Also Serve, As An Ideal, Induction Free Energy Platform, A Clamped Double Saucer Configuration, Perfectly Configured, To Catch And Modulate, Lagrangian Curvatures, Of Gravitationally Modulated, Magnetic Tensor Waves, Created During Hyperacceleration, Quantum Leap Events - The Faster The Acceleration Rate, Of The Hull/Fuselage, Ballistic Projection, The Greater, The Co-Efficient, Of Zero Point Free Energy Induction, Is Commodified, Into Surplus Propulsive Forward Momentum - Leading To Quantum Leap, Past The Speed Of Light, In Vacuo

145) The Periodic Table Of Elements, Ultra Heavy Isotopes, With Atomic Numbers, In The Post Uranium Scale, In This Complex, And Ultra Dense, Molecular Atomic/Number Range: Are By Nature, Synergistically Entwined, When Drawn Into, Interactive Physics, In The Planck, Fusion Nuclear, Power Factor Range: The "Planckean", Infinite "Gravitation Transduction", Through Tensor Rotation, Creating Gravitationally Generated Autonomous (Anti-Gravitational, Unique/Exotic Events) In Fact, Synthetically Identical To A, Super-Compressed, Black Star - Dark Energy Induction, Lagrangian Modulated, Plasma Field, With Stabilized Tensor Spins, In "Homeostatic Centrifuge": Magnetically,

“Torsion Regulated”, From Sidereal Sway, And Wobble, Under Tremendous, Unilateral, Unidirectional/Symmetrical, Gravitational Force Field - Micro-Correcting, The General Containment Field, From “Free Chaotic” Tensor Asymmetrical Spin, En Vacuo -
146) Within, The Ever Expanding, Periodic Table Of Elements, Atomic Weight Numbers, In The Atomic Number Range, Of 115 To 119, Seeming To Be, The Paradigmatic Ideal, @ Current Known Technology Levels, Of “Synergistic Interactivity”, Against Radioactive/Fissile Bombardment, And Faraday Isolation/Protection - Black Hole Suns, Transduced, From Gaseous Incendiary Plasma, To Super-Dense, Gravitronic, Induction Magneticism, Must, In All, Probability, Share Similar Atomic, Periodic Table, Super Molecular Structure, “Dense Configurations”, In An Escalating Scale, Of Molecular Complexity/Density/Energy Output: Requiring Minimal Rotation, Due, To Ultra Density: To Generate Black Hole, Super Gravitational Induction - Ultra Gravitational Density, Mass Compressed, Super Magnetic Spheroids: Are Still Able, At An Infinitesimal Fraction, Of Their Original, Incendiary Gaseous, Solar Mass: To Physically, Magnetically, Tether/Retain, Their Original “Solar Heliosphere”, Indicating, “Identical Solar, Amalgamated Weight Mass”: Additionally Creating, In Such, A Magnetic “Phantom Heliosphere”, An Anti-Matter, Solar Stargate, “Induction Conduit”, Sufficient To Draw, Any And All Forms, Of Localized Energy, Into Itself: While Simultaneously, Emitting, Ultra Velocity Quantum Particles, Consistent, With Anti-Matter, Interdimentional, Worm Hole. Spontaneous Formation = “Downstepped”, Momentarily, From Hyperspace, By The “Braking” Tethering, Massive Gravitational Pull, Of The Black Sun, “Worm Hole” - Once Sufficient Distance, Is Achieved, The Solar Magnetic Influence “Weakened”, The Quantum Particles, Slingshots, Back Into Hyperspace, Demonstrating “Virtual”, Material/Energy, Propulsion “Exchange”, Unique Physical Paradoxes, In The Material/Energy, “Decay Process” Once Distanced, From The “Magnetic Bottle”, Of The Black Hole, “Stabilizer Field” - This, “Unified Field Theory”, Can Be, Technologically Exploited, Mimetically Synthesized: To Create, A Quantum Leap, Warp Event, In Current Physics: Implying Deep Space Travel, Free Energy Generation, Terra Forming, Planet Space Colonization, Creating Synthetic Suns, Satellite Biospheres, Synthetic Toroid Worm Hole Sets, Terrestrial Weather/Natural Disaster Control, And Ultimately, Level One Civilization: In Effect, Creating A New Age Paradigm Shift, On A Total, Universally Beneficial Scale, For The Progress Of Mankind - Solar Physics, Provides, The Essential Scientific Key, And Technological Solution: And By It’s Physical Study, We Find The Mystery, Of Genesis Itself, Partially Revealed: From The Most Volatile, And Combustable, Simple Molecular Gases, Transforming Through Time, Evolving Through, The Atomic Gauntlet, Of The Periodic Table, To The Most Dense Magnetic Material, Theoretically Known, Constantly Escalating, In Energetic Output, And Productive Inductive/Productive, Output Per Mass Valence, As It Evolves, From A Solar Celestial Body, Into An Interdimensional Portal, And Worm Hole Nexus - Demonstrating, In Solar Evolution, The Cause And Effect, Of Unified Field Physics
147) What Will Become Of Us: Within The Lagrangian Tensor, Spiral Epicenter, Of The Free Energy Construct, Constellated Mass Frame: Black Hole, Convening, Di-Opposed, Electro-Magnetic, Twin Whirlpool Siphons, Of Valence Charged, “Fictitious Fluid” Dynamics, In The, “Poincare Conception”, Of Electro-Physical E/M, A/G, Fields, Maintains, Stabilized “Centrality, Within The “Mass Frame”, Central Torquing, Axis Of Rotation, Solar Physics Fusion Modulus: Not Unlike, The Windless Calm, Within The

Eye, Of A Hurricane, Exerting, Tugging, Trajectory Influence: Like A Gear/Spline, Screw Rotated, Motor Armature, Or A Combustion Engine, Piston Rod Array: The Axis Of Rotation, Becoming Irrelevant, As Time Space, Is Scalar Graduated, And Eliminated, In Simultaneity: Vectors Of Euclidean Rotation, At Hyper Spun Tensor Rates, Amalgamate Into, A Holographic, Stasis Plane Rate, In Perpetual, Exotic, Physical Warping Event: As It Transits, The Entwined External Event = Entrance/Exit Horizons, Central Black Hole Vortex, And White Hole Terminus, Exit Point, In Simultaneous, Linear Trajectory - Through The E/M Generator Conditioned, Poincare Topology: Of Macrocosmic To Microcosmic, Interdimensional Gradations, Of Siphon Funneled Energy, Upon A, Three Dimensional, Then Four Dimensional, Energy Focusing, "Coiled Gradient": Upon Sequentially Truncating, Spheroid Planes, Reaching Maximum Reducibility, In Space Time Reality: At The Tensor Energy Stabilized, "Yoke Point" Of The Black Hole, Induction Nexus: Invariably Existing, At The "Rotational Center", Of The, "General Field Reaction" - Interdimensional Time Travel, Thus Occurs, Through This Eye, Of The "Celestial Needle" - In Stellar Cosmological, Astrophysical Terms - Ubiquitously, Generating And Distributing Free Energy, Through An Infinitely Expanding Universe, Turning Matter Into Energy, Through Acceleration, Then Downstepping Velocity, Into Normative Gravitational Equilibrium, To Form Celestial Bodies, Which Are General Field, Enmeshed, Through Electro Magnetic, Orbital, Gyro-Rotational, Galactic, Heliocentric, Planetary, Lunar Influence: All The Way Through, To The Subatomic, Virtual Particle/ Dark Matter, Isotopic Level - Beginning And Ending, As It Were: At The Amniotic/ Osmotic Level, Double Mirror, Co-Valence, Set Helix, White And Black, Hole Pairings: Which Exist Everywhere And Nowhere, In Bilocation Simultaneity, Always And At Once
148) The Topological Surface, Of Our Heliocentric Sun, At 6,700 Degrees, Fahrenheit, Is Consistent, With Exothermic-Dynamic Yields, Of Atmospheric Fission Nuclear Explosions, As Well As Carbureted, Fire Bombing, And Wind Factor, "Blast Furnace" Effects, Upon War Targeted, Urban Street Grids - These Intense, Thermo-Nuclear Level, Heat Factor Yields, Were Not Unknown, To The Sumerian/Babylonian Civilization, Who Had Developed Metallurgical/Alchemical Furnaces, Sufficient, To Create Super-Heavy Isotopes, 115-119, From Calcium Oxide, And Gold, Allegedly, Used To "Fuel" And "Open" Interdimensional, "Stargates" In An Around, Their Ziggurat Pyramid Centered Cities, In The Tigris-Euphrates, Fertile Crescent, River Confluence Industrial Thermite, Chemical Ignition, Begins @ 4,000 Degrees F, And Synergistically Climbs, Thermodynamically, As It Turns Solid Gold, Molten, W/Periodic Atomic Number 79, And Reduces, Steel Reinforced, Structural Rebar, To Molten Slag And Slurry, Within Caste Concrete: Which In Turn Becomes, Powderized, Calcium Oxide, Periodic Atomic Number 40 = Element 119 = When All Moisture, In The Concrete, Is Vaporized, Trapped Air Pockets, Expand And Explode, And Heat Twisting Rebar, Acts, As A Levered Prybar: The Constituent Parts, Forming A Super-Isotope-Element-Molecule - As In The Case, Of The Blast Furnace, Of Nebuchadnezzar, Dedicated To Bel(ial) And Various, Of The Anunnaki "Pantheon" Of Nefilim Giants - Which Were Used For Holocaustic/ Crematoria, Human Sacrifice, To Their Extraterrestrial Gods - Human Skeletons = Calcium Oxide, Atomic Number 40 + Pure Gold - Concrete Unavailable, An Invention, Of The Etruscan/Romans - To Modern Day Catastrophes Like 9/11, Where A Large Horde Of Gold Bullion, Twin Skyscrapers Built Of Concrete, And 2,000+ Incinerated Victims, Were Apparently, Destroyed, Murdered, Sacrificed?, In A 4,000+ F, Towering

Inferno: Allegedly Caused, By Dual Jet Collision, And Subsequent Spillage, Then Ignited Rocket Fuel: With A Burning/Combustion Temperature Of, 1,017 Degrees F, Being Primarily Mixed, With **Kerosine** - One Can Readily See "The Problem", In The Alleged Thermodynamics, Of The Event - Apparently Thermite, In Substantial Quantity, Or A Mini-Nuclear Devise Was Involved, To Cause A Nearly Perfect, Gravity Rate Fall, Of The Structures, Consistent, With An Engineered Building Demolition -

149) However Tragic, These "Events" That Have Occurred, Throughout Human History, Where Apparent "Contact" **1 Thru 6**, Has Been Made, With Advanced Intelligence, And Their Related Technology, "Interface": It Is Therefore, Of Interest, At This Point, In Modern Civilization, Where Human Science, Can Now Forensically, "Deconstruct, The Physical Evidence", The Reason, For Uniting, These Materials Together, Under Extreme Heat, And Catalytic, "Molecule Into Element", Maximized Level, "Atomic Binding Pressure", To Create Super Elements, From Hoarded Precious Metals, And Calcium Compounds - We Come To Realize, There Is A Method, To The "Madness Of Nebuchadnezzar", In The Biblical Sense, In That, The Produced Material, When Cast Founded, Into Molds And Sheets, Can Form, Nearly Impregnable Fuselage Components, For Deep Space Vehicles - And The Earth, Has A Vast Abundance, Of Both Materials Required, In Question - We Then Realize, There Is Nothing New, Under The Sun, And We Are, In Effect, Retro-Engineering Ancient, Extraterrestrial Technology

150) Even With, The Tensile Strength, And Molecular Density Protection, Of A Super Post Uranium Element, Space Vehicle, Hull Construction - Deep Space, Gamma, X-Ray, Solar Radiation, Over Time: Will **Always,** Be A Deleterious Health Factor, Therefore, It Is An Imperative: To Decrease Relative, Travel Distance: Through Black Hole, White Hole, Fusion Powered, Tele-Portation, Spacio-Temporal Displacement, Production - It Is Also Essential, To Truncate, Deep Space Travel Time, And Eliminate, Light Year Distance, Through Interdimensional, Quantum Leaps - Emulating And Synthesizing, The Unified Field Fusion Physics: Of Black Hole Suns, And The White Hole, Twin Star Gate Sets - Perhaps, Through Future Exploration, Finding Remnants/ Remains, Of Analogous Technology - Once We Know, What We Are Looking For -

151) The Interdimenional, "Contigouous, Atomic Amniocentesis" Of Matter/Anti-Matter Exchange: Across Diaphragmatic, Omnidirectional Hyper-Acceleration, Warp Field Distribution, Within The Stablilized Black Hole, Flux State Ambiguity, Of Both Portal And Barrier - Requires, A Four Dimensional Trigonometry Matrix: Gravity/Anti-Gravity, Rate Table, As Well As Second Law, Of Thermodynamic, Reconsideration/Addendum/ Revision: Factoring In, Not Only, The Environmental Energy/Temperature, Acceleration Work Conversion, Of Material Trajectory , Beyond Alleged Terminus, Of The Light Barrier - But Also Considering, The Amalgamating, Graduated Rate, Of Truncating Spin Tensor Dynamics, Within The Di-Opposed, Magnetic Excitation, Conical Whirlpooling Fields: Creating, In Effect, A "Phantom" Caducean Toroid Wound, Fermi/Einstein Cyclotron, Induction Focusing Mag/Lev Coil, Which Effects Matter Weight Distribution, Along Isotopic Tow Lines, Of Lagrangian Force, Upon A Linear Ballistic Acceleration, Of I/D Portal Induction - Like An Arrow Shot Downward, Through The Eye Of A Hurricane, Or A Jetplane, Nosediving, Towards Earth Gravity, In Severe Angularity, To Create Artificial /Temporary (Unstable Tensor Field) Cabin "Weightlessness", This Computational/Topological Model, Of Relativity/Simultaneity, Field Dynamics Interplay, Would Help Reconcile, A Mathematic/Theoretical Framework, For Unified Field Fusion

Dynamics, Interplay, And The Coinciding, Relativistic Physical Field Events, That Occur Around Them: As Exotic, Discreet, And Chaotic Nuclear, Temporal Displacement Phenomena - The Mathematic Model, When Converted, Into A Working Conceptual Model, Would In All Probability, Resemble, And Or, Improve Upon, Current Cold Fusion Tokamak, Breeder Reactor/Generator, Nuclear Engineering/Industrial/Defense/ Aerospace Design: Eventually Leading, To Perpetual Warp Engine: Essentially, Creating Solar Physics, Void Of Thermodymamic Breakdown, In Locomotive Perpetuity - Solar Forming, And Terra Forming, Near/Deep Space Colonization, Would Be The Next Logical, Scientific Step, To The Ultimate Evolutionary Goal, Of Authentic, "Level One Civilization": The Control Of Time Space, Catastrophic Weather, And Sky Defense

152) Another Useful Demonstration, Of Simplifying By Illustration, Such An Magnetic Energy, Tensor Flux Capacitance, I/D, A/G, Fusion Matrix: Would To Visualize, A Twin Set Of Antique, Meat, Or Coffee Grinders, Welded Together At The Extrusion Ports, To Form A Single Tensor Ring (Black Hole) The Ground Material Is Being Leveraged, Truncated Down, By Raw Material "Feeding Pressure" Into The Two Devices - The Hyper Spinning (Tensor Lagrangian Magnetic, Topological Conic Field) Of The Grinding Crank, Creates Additional Inducted Energy/Work/Pressure, From The Surrounding Environment = (Clausius/Helmholtz/Maxwell/"Thermodynamic/Entropic" Induction Work Surplus, Via., Lorentzean, E/M, A/G, "Self Relieving", Gravitational Power Factor) Due To The "Warp Phase Shift", Hyper Acceleration, Of The Introduced/Inducted Matter, (Ground Meat, Coffee, Coming From, "Competing Field" Opposite Directions, Of The Siamese Twinned, And Welded Together, Antique "Grinders") The Anti/Phantom Matter, Dark Energy, Virtual Matter, In Question, In Order To Flow, And Not Destroy, The Joined "Apparatus" With Nuclear, "Back Pressure" Must Pass "Through Itself"/ Passing Through The Incoming, Virtual/Warped Material, In Order To Exit, Into Three Dimension "Form", Downstepped "Normal Field" Accelerated Space - While Transporting It's Energy, Plasmatically, Through General Field Distribution, To Relativistic Space - By Thus Discharging, It's Materially Conditioned, "Temporary Valence, 3D State", It Spontaneously Reaccelerates, Back To The Other Side, Of The Interdimensional/ Siamese Twin, "Meat Grinder" And Continues, The Exact Same, Nuclear Process, Perpetually, This "Valence Reaction/Matrix Outcome" Can Neither Be Created, Nor Destroyed, In This Cyclical Manner - However Expansion And Acceleration, Are Fueled In This Manner, In Thirty Billion Year, Big Bang/Extension Bust Cycles, When The Amalgamation Of All Known Matter, Hyper-Accelerates Beyond The Light Barrier, Subsequently, Big Banging Again, In Another Thirty Billion Year Cycle, And So Forth

153) Like The Mythological Example, Of Typhon (= Typhoon), Most Powerful, Wind Water/Fire Demon, Of The Vanquished, Greek Gigantomachy: Ruthless Son, Of Gaia, The Earth Mother - Transfixed, By The Thunder Bolts, Of Olympian Zeus, For Warring Against Him, With Incited, Chaos Rebellion, Of The Elements: The Giant Typhon, Sought To Dominate, The Vast Space Cosmogony, Through Controlling Natural Forces, To His Iron Willed, Advantage - Now Pinned Beneath, The Twin Conical, Volcanic Mountains, Of Aetna, And Vesuvias: "Dragon Belching", Raging Fire, From Beneath, Molten Lava Forming, Heating, The Cyclopian Smithy, Of Vulcan - To Hammer Forge Wrought, Impervious Heaven Metal (Element 115-119) = The Myth Explained, In More Scientific, Modern Jargon, In Terms Of A Thermodynamic, Lagrangian/Lorentzian, "Typhoon/Whirlpool", Two Energetic "Energy Cones" = "Phantom Mountains", Valence

Yield Output/Synergistically Combined By Induction = Cojoined, By Contiguous, Diaphragmatic Space Portal, “Siamese Twin, Interdimensional Interface” Pinned Down, And “Lightning Transfixed”, (Matter, Linear Accelerated, Through Centrifuge/Gyroscopic Induction: Straight Through, To The Black Hole, Precise Center, Of The Unified Fusion Field, Nuclear Reaction = (Vulcan Blast Forge/Solar Furnace) Aided By His Cyclopian Apprentices = Symbolic Of The Myopic/Focused/Powerful (“One Eyed”, Black Hole Sun)

154) Hubble “Entropic Flow” Away, From Earth Gravitational Constraint, Upon Angstrom Cosmological Light Unit, Of Photon Acceleration, Is Empirically Observed To Be Linear In Nature, And Geometrically Progressive, In Acceleration Characteristic: Implying Valence Increase, By Flight Path Induction, And Magnetic Field Gravitational Formation Into Dark Energy Quantum Field Energy Variables, Upon Surrounding Interlaced Trajectory Fields, Being Drawn Into, With Matter Manipulated, By Compounding Gravitational Confluence, Creating An Inductive Ambient, Synergistic Entropic, Exponential Yield Potential

155) By Extrapolation/Implication, Like The Ballistic Force, Of A Bullet, Striking A Target, At High Acceleration, Mass Impact, Increased By Centrifugal Penetration, The Lagrangian Plume, That Forms Around, The Intergravitational, Super Light Speed, Quantum Event: Inversely Proportionate, To Terrestrial Distance, From The Hubble Constant - Creates, An Overlapping Cone Matrix, Not Unlike A Dense Stone, Spun Cast, Then Entering, An Subsequently Agitated Plasma Field, Of Once Tranquil Pond Water - This In Turn, Is Mirrored, Upon The Obverse Side, Of The “Lagrangean Penetration Plume” In Causial Simultaneity - As The Same Matter Exists, Simultaneously, In Dual Spectra, Of The Time Space “Displacement Quantum” Co-Valence, Phenomena, “Nuclear Fusion Event Set” - Like Two Sides Of An Ink Blot

156) The Actual Time Barrier Event, In Three Dimensional Space, Acts, Both As The Tranquil Pond, Calm, Undisturbed Surface, And A “Magic Mirror”, About To Become, Four Dimensionally, “Osmotic”, And Four Dimensional Portal, To HyperAccelerated Light No Longer Acting, As A, “Reflective Barrier”, But As An Interdimensional, Linear Light, Black Hole, “Tensor Yoke Portal”, Accommodating And Focusing, The Transdimensional Light, Turned Into Energy, By Thrust Projection, Into Quantum Leap - Space And Time Are Thus Neutralized, Within The Linear Quantum, “Simultaneity Path” - Structural Integrity, Of The Rematerialized Matter, Contingent Upon, The Neutralization, Of Time Space Influencing, Three Dimensional Surrounding Quanta @ “Specialized Cosmology”

157) Unimpeded Hubble, Linear Photon Acceleration, Including Induction Torque, At Inter Dimensional, “Amalgamated Quantum, Leap Point”, In Tandem Relation, To Transported Leap Mass, Assures Homogenous, Reaction Symmetry -

158) Spacial Distance, Truncated At, “Zero Point, Event Horizon”, At The Centrific Core Of The Reaction - Creates a Lagrangian, “Magnetic Rifling Effect”, Of Projectile Entry Ballistic Path, Like A, “Caduceus Coil, Fermi Cyclotron” - Although Linear, And Tornado, Like, In Dynamic Structure - Truing The Alignment, And Tensor Spin, To Optimal Super Spun, Spiral Stability - Further Ensuring, Navigationally Accurate, Rematerialized Target Destination/ Time Point, When Sufficient, Valence Deceleration, And Required Distance, Has Been Established, From The Star Gate Portal: Omnidirectionally “Escaping”, Euclidean Matrix Black And White Hole, Energetically, “Siamese Twinned” Productive Termini - The Star Gate Reversing Gravity, Serves To Decelerate And Help Materialize The Energized Matter: If Sufficiently Dense, It Will Escape Re-Entry Into The Black Hole

A) Quantum Leap, Geometrical Acceleration Curve, Time Warping Conical Architecture, Spiral Configuration
B) "Hubble" Linear Light Trajectory, Of Photon/Angstrom Units:
C) Through "Lagrangian Rifling" Through Magnetic, Geo/Solar Axis Spiral:
Torque Rotation, "Focusing Coil" Inducting To "Eternal Yoke Tensor"
D) Black Hole, Stargate Portal, "Event Horizon" Outer Influence Perimeter, Of Spacio-Temporal Event
E) Time Anomalous, "Special Events", In Proximally, "Effected Space", Proportional To Magnitude, Of Black Hole Power Factor Reaction, Causing Dimensional, Virtual Flux
F) Magnetic/Levitation/Black Hole "Time Collapsed, Field Tensor" Wave Form Ballistic, Energy Dynamic Plume, "Plasmoid Modulus": Reaction Containment Circulation System
G) Horizon Trajectory, Warp Factor Point: Pi = Warp Interdimension, Into 4D Phasing
H) Lorentzian, Magnetic, Cone/Yoke/Tensor: Siamese Twinned, Black Hole: 2X Cone Centrifugal, "Central Reaction Connector"
I) Tachyon Particle Flow, From Transverse, Omnidirectional, Centralized Reaction Point, Emanating, From White And Black Hole, Event Horizons, Into Relativistic, 3D Effected "Special/Exotic Event" Environmental Reaction In Surrounding Space ("Ambient Stator")
J) "Fermi" Caducean, Wound/Counterwound, Plasma Coil Cyclotronic, Focusing Coil, Magnetic Wave, Lagrangian, Grid Matrix: 2X Siamese Counterspun, Plasma Cone Coil Stargate Sets, Matched Mass, Energy Valence, Rotationally Counterspun, With Central Crushing Propulsive Force, Drilling Into The Tensor Ring, To Warp Tear, Into 4D Space
K) "Plasma Whirlpool" Siamese Twin, Bell Shaped Curved Cones, Which Expand From The Center Nexus, To The Energy Plumes, Which Act As, "Gravitational Flyballs", Like "Governors", Found On Steam Engines, Which "Calibrate" Rotation, As Well As Help Intensify Acceleration, By Increasing, Rotational End Point, Spin Baring, Of Mass Load
L) Hubble Linear Acceleration Trajectory, De-Acceleration Point, Rematerializing After Teleportation Beyond, The Horizon Event Corona, Of Either Side, Of The Black Hole When Sufficient Distance Is Established, To Eliminate Inductive, Towing Pull Reaction
M) Golden Section, Counterspun, Spiral Modulus, Of Siamese Twinned, Bell Cone, Black Hole Configuration, Is Formulated, By Internal And External, Power Plasma Dynamics, Magnetically Energized, Through Hyper-Rotation, And Kept From Structural Collapse, Of The "4D Eternal Tensor Yoke", By External, 3D Environmental "Containment", And Internal Induction/2X Cone "Inflation", Of Hyper-Rotated, Free Energy, Fueling The Interdimensional, Reaction Core: Opening And Maintaining, The Stargate 4D Portal, Creating, A Homeostatic, Isotonic Tensioned, Yet Ultra Dynamic, Spiraling Structure: The Two Siamese Twinned, Plasma Cones Compressing, "Head To Head", At Incredible Counterspun, Colliding Propulsive Pressure: They, In Turn Create, A Dimensional Warp, And In So Doing, Complete, The Missing Mathematic Solution Factor, Of Pi, Which Exists, At Their 4D Interdimensional, Fusion Point Interface, @ The Black Hole Stargate, Is In Fact, The Elusive Interdimensional, "Missing Numerator" Within, @ The Crushed Epicenter, Of The "Eternal Tensor", "Dual Concave", Free Energy/Dark Matter, Induction Energized, Interdimensional 4D, Spacio-Temporal Displacement, Nuclear Fusion, Teleportation Device - In Effect, Short Cutting, Spacial/ Time Distance, To Absolute Zero Point, As Surrounding, Displaced Space, Interval Involved, Is Warped And Folded, Into Itself: In Coexisting Simultaneity, At Both Black/ White Hole Reaction, Event Horizon Termini, Instantaneously, Eliminating Relativity

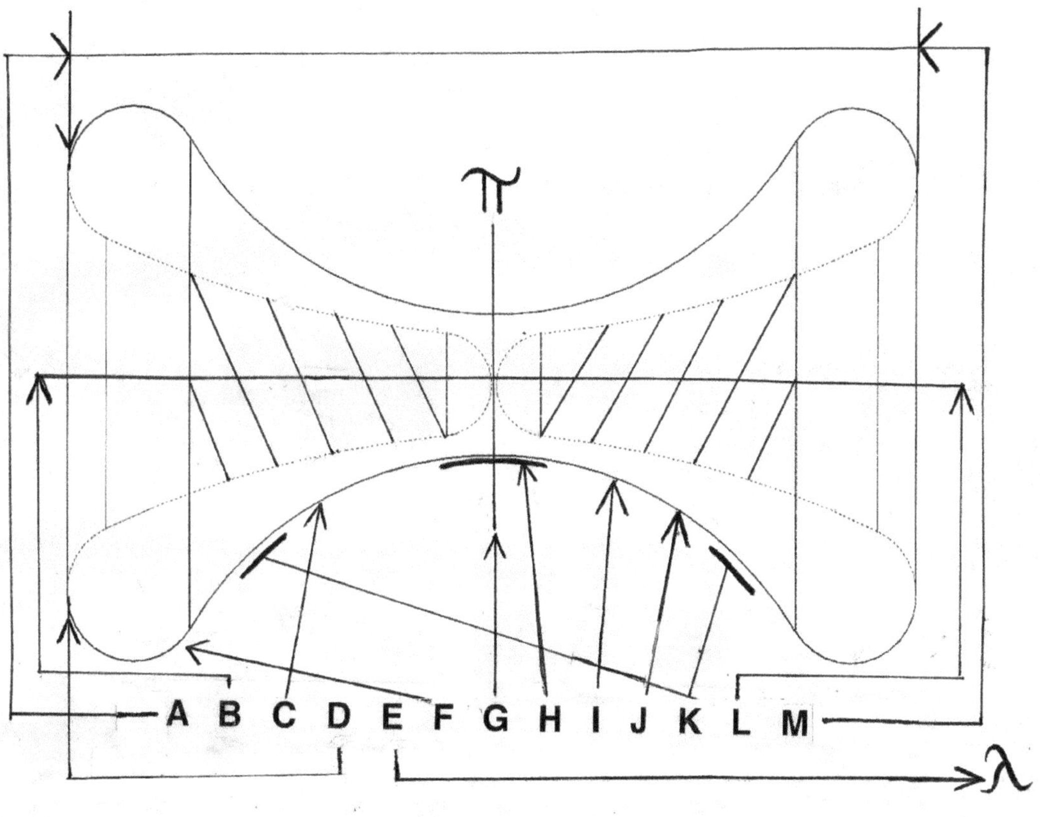
π
A B C D E F G H I J K L M
λ

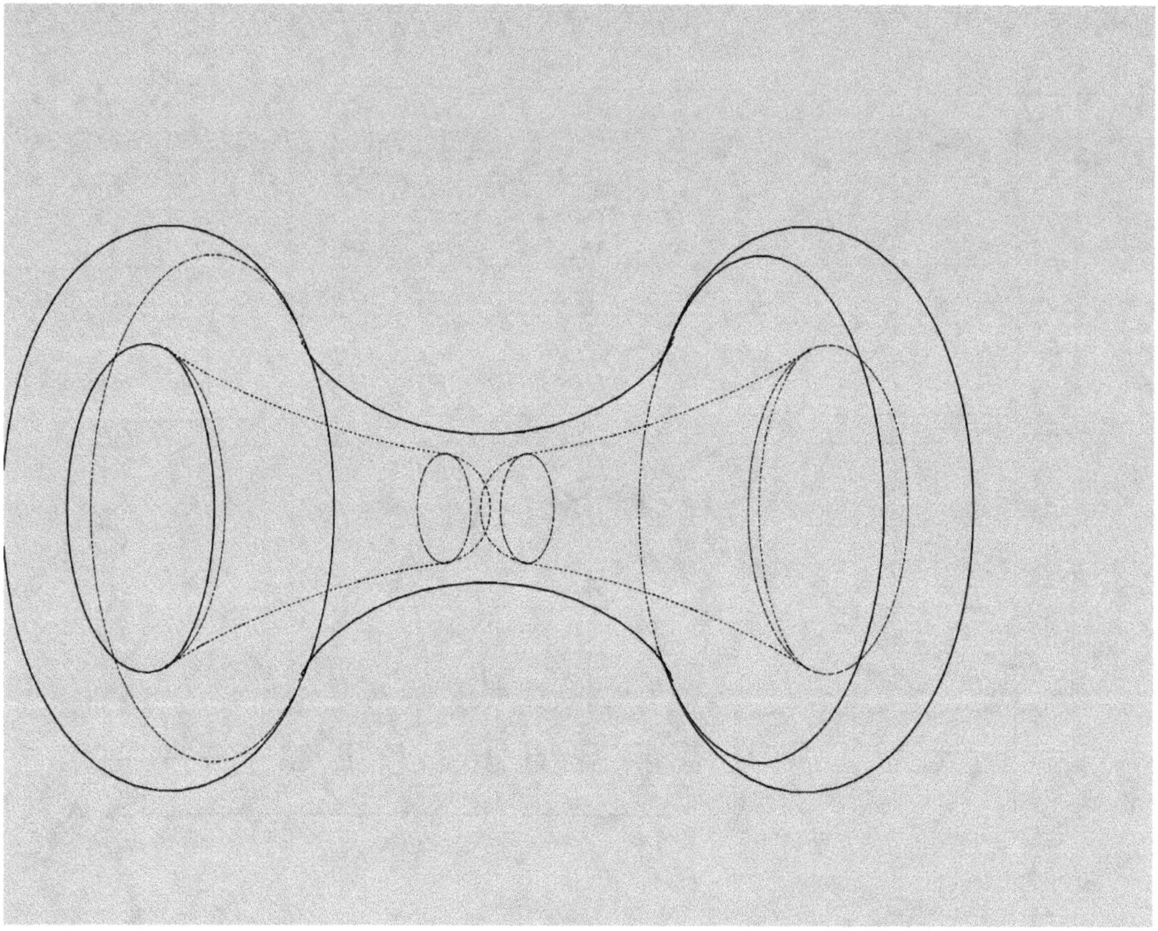

Octalinear Anti-Matter Catalyzed Impulse Fission/ Fusion Reactor

Octalinear

dimension:

project:

Ver: 2

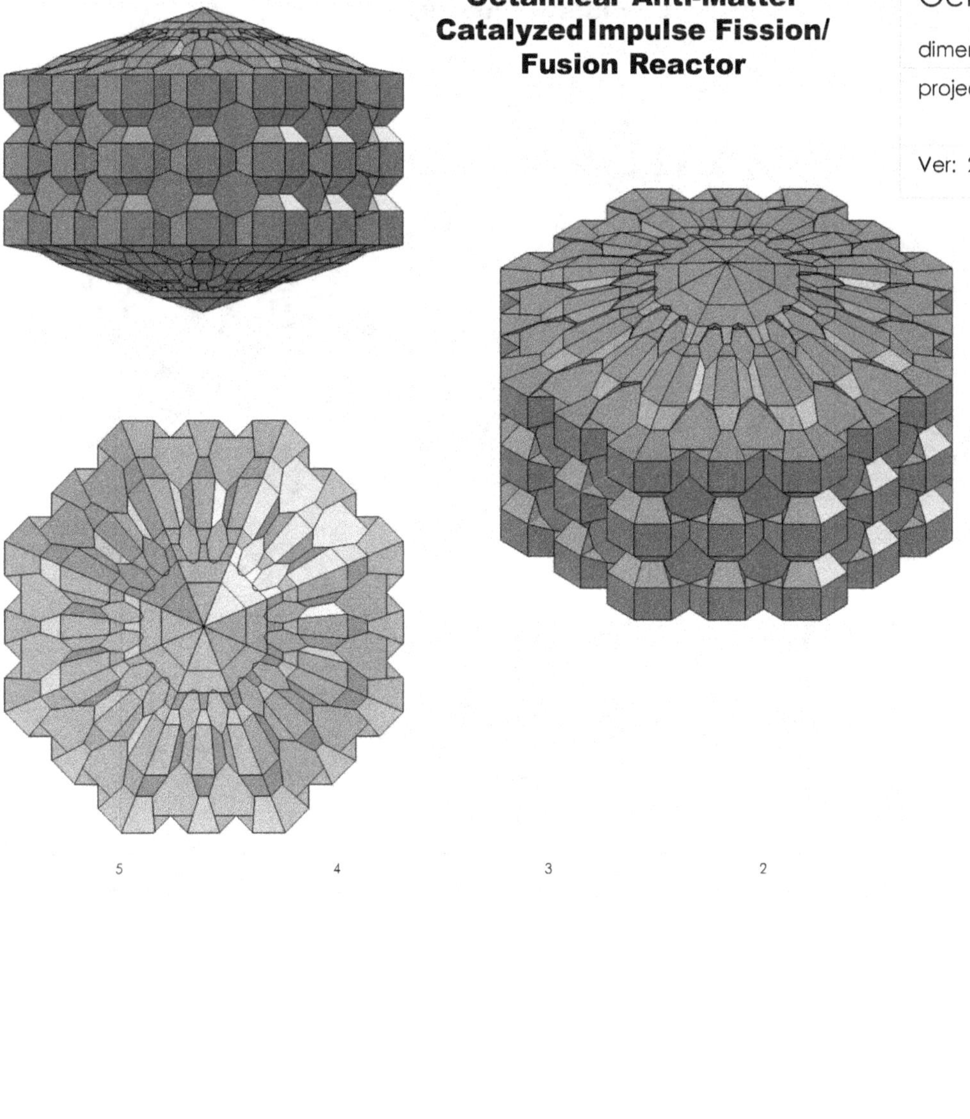

5 4 3 2 1

Octalinear Anti-Matter Catalyzed Impulse Fission/Fusion Reactor

www.ingramcontent.com/pod-product-compliance
Lightning Source LLC
Chambersburg PA
CBHW080823170526
45158CB00009B/2512